APPLIED PHARMACEUTICAL PRACTICE AND NUTRACEUTICALS

Natural Product Development

APPLIED PHARMACEUTICAL PRACTICE AND NUTRACEUTICALS

Natural Product Development

Edited by

Debarshi Kar Mahapatra, PhD

Cristóbal Noé Aguilar, PhD

A. K. Haghi, PhD

First edition published 2021

Apple Academic Press Inc.
1265 Goldenrod Circle, NE,
Palm Bay, FL 32905 USA

4164 Lakeshore Road, Burlington,
ON, L7L 1A4 Canada

CRC Press
6000 Broken Sound Parkway NW,
Suite 300, Boca Raton, FL 33487-2742 USA

4 Park Square, Milton Park,
Abingdon, Oxon OX14 4RN

First issued in paperback 2023

Apple Academic Press exclusively co-publishes with CRC Press, an imprint of Taylor & Francis Group, LLC

Publisher's Note
The publisher has gone to great lengths to ensure the quality of this reprint but points out that some imperfections in the original copies may be apparent.

Library and Archives Canada Cataloguing in Publication

Title: Applied pharmaceutical practice and nutraceuticals : Natural Product Development / edited by Debarshi Kar Mahapatra, PhD, Cristóbal Noé Aguilar, PhD, A.K. Haghi, PhD.

Names: Mahapatra, Debarshi Kar, editor. | Aguilar, Cristóbal Noé, editor. | Haghi, A. K., editor.

Description: First edition. | Includes bibliographical references and index.

Identifiers: Canadiana (print) 20200367420 | Canadiana (ebook) 20200367528 | ISBN 9781771889247 (hardcover) | ISBN 9781003054894 (PDF)

Subjects: LCSH: Drug development. | LCSH: Natural products.

Classification: LCC RM301.25 .A67 2021 | DDC 615.1/9—dc23

Library of Congress Cataloging-in-Publication Data

CIP data on file with US Library of Congress

ISBN: 978-1-77188-924-7 (hbk)
ISBN: 978-1-77463-776-0 (pbk)
ISBN: 978-1-00305-489-4 (ebk)

DOI: 10.1201/9781003054894

About the Editors

Debarshi Kar Mahapatra, PhD
Assistant Professor, Department of Pharmaceutical Chemistry, Dadasaheb Balpande College of Pharmacy, Rashtrasant Tukadoji Maharaj Nagpur University, Nagpur, Maharashtra, India

Debarshi Kar Mahapatra, PhD, is currently Assistant Professor in the Department of Pharmaceutical Chemistry at Dadasaheb Balpande College of Pharmacy, Rashtrasant Tukadoji Maharaj Nagpur University, Nagpur, Maharashtra, India. He was formerly Assistant Professor in the Department of Pharmaceutical Chemistry, Kamla Nehru College of Pharmacy, RTM Nagpur University, Nagpur, India. He has taught medicinal and computational chemistry at both the undergraduate and postgraduate levels and has mentored students in their various research projects. His area of interest includes computer-assisted rational designing and synthesis of low molecular weight ligands against druggable targets, drug delivery systems, and optimization of unconventional formulations. He has published research, book chapters, reviews, and case studies in various reputed journals and has presented his work at several international platforms, for which he has received several awards by a number of bodies. He has also authored the book titled *Drug Design*. Presently, he is serving as a reviewer and editorial board member for several journals of international repute. He is a member of a number of professional and scientific societies, such as the International Society for Infectious Diseases (ISID), the International Science Congress Association (ISCA), and ISEI.

Cristóbal Noé Aguilar, PhD
Director of Research and Postgraduate Programs at Universidad Autonoma de Coahuila, Mexico

Cristóbal Noé Aguilar, PhD, is Director of Research and Postgraduate Programs at the Universidad Autonoma de Coahuila, Mexico. Dr. Aguilar has published more than 160 papers in indexed journals, more than 40 articles in Mexican journals, and 250 contributions in scientific meetings.

He has also published many book chapters, several Mexican books, four editions of international books, and more. He has been awarded several prizes and awards, the most important of which are the National Prize of Research 2010 from the Mexican Academy of Sciences; the Prize "Carlos Casas Campillo 2008" from the Mexican Society of Biotechnology and Bioengineering; National Prize AgroBio–2005; and the Mexican Prize in Food Science and Technology. Dr. Aguilar is a member of the Mexican Academy of Science, the International Bioprocessing Association, Mexican Academy of Sciences, Mexican Society for Biotechnology and Bioengineering, and the Mexican Association for Food Science and Biotechnology. He has developed more than 21 research projects, including six international exchange projects. His PhD in Fermentation Biotechnology was awarded by the Universidad Autónoma Metropolitana, Mexico.

A. K. Haghi, PhD
Professor Emeritus of Engineering Sciences, Former Editor-in-Chief, International Journal of Chemoinformatics and Chemical Engineering and Polymers Research Journal; Member, Canadian Research and Development Center of Sciences and Culture

A. K. Haghi, PhD, is the author and editor of 200 books, as well as 1000 published papers in various journals and conference proceedings. Dr. Haghi has received several grants, consulted for a number of major corporations, and is a frequent speaker to national and international audiences. Since 1983, he served as professor at several universities. He is former Editor-in-Chief of the *International Journal of Chemoinformatics and Chemical Engineering* and *Polymers Research Journal* and is on the editorial boards of many international journals. He is also a member of the Canadian Research and Development Center of Sciences and Cultures (CRDCSC), Montreal, Quebec, Canada.

Contents

Contributors

E. D. Ahire
Department of Pharmaceutics, Sandip Institute of Pharmaceutical Sciences, Nashik, Maharashtra, India

Mojabir Hussen Ansari
Department of Pharmaceutics, Gurunanak College of Pharmacy and Technical Institute, Nagpur 440026, India

Vivek Asati
Department of Pharmaceutical Chemistry, NRI Institute of Pharmacy, Bhopal 462021, India

Akshada Atul Bakliwal
Department of Pharmaceutics, Sandip Institute of Pharmaceutical Sciences, Nashik, India

Taranpreet Kaur Bamrah
Department of Pharmaceutics, Gurunanak College of Pharmacy and Technical Institute, Nagpur 440026, India

Sanjay Kumar Bharti
Institute of Pharmaceutical Sciences, Guru Ghasidas Vishwavidyalaya (A Central University), Bilaspur 495009, India

Pooja Bhomle
Department of Pharmaceutics, Gurunanak College of Pharmacy and Technical Institute, Nagpur 440026, India

Vijay Sharadkumar Chudiwal
Department of Pharmaceutics, Sandip Institute of Pharmaceutical Sciences, Nashik, India

Kanhaiya M. Dadure
Department of Chemistry, J. B. College of Science, Wardha 442001, India

Shruti Dongare
Department of Pharmaceutics, Gurunanak College of Pharmacy and Technical Institute, Nagpur 440026, India

Sónia I. G. Fangaia
Faculty of Medicine, University of Coimbra, Av. Bissaya Barreto, Blocos de Celas, Coimbra 3000, Portugal

Fernando A. D. R. A. Guerra
Faculty of Medicine, University of Coimbra, Av. Bissaya Barreto, Blocos de Celas, Coimbra 3000, Portugal

Dipak Kumar Gupta
Department of Pharmaceutics, School of Pharmaceutical Education and Research, Jamia Hamdard, Hamdard Nagar, New Delhi 110062, India

Animeshchandra G. M. Haldar
Department of Applied Chemistry, Priyadarshini Bhagwati College of Engineering, Nagpur 440024, India

Sameer A. Hedaoo
Department of Pharmaceutics, Gurunanak College of Pharmacy and Technical Institute, Nagpur 440026, India

Paras Kothari
Department of Pharmaceutics, Gurunanak College of Pharmacy and Technical Institute, Nagpur 440026, India

V. M. M. Lobo
Department of Chemistry, University of Coimbra, Coimbra 3004-535, Portugal

Debarshi Kar Mahapatra
Department of Pharmaceutical Chemistry, Dadasaheb Balpande College of Pharmacy, Nagpur 440037, India

Pedro M. G. Nicolau
Faculty of Medicine, University of Coimbra, Av. Bissaya Barreto, Blocos de Celas, Coimbra 3000, Portugal

Vaibhav Shende
Department of Pharmaceutics, Gurunanak College of Pharmacy and Technical Institute, Nagpur 440026, India

Vijayraj N. Sonawane
Department of Pharmaceutics, Sandip Institute of Pharmaceutical Sciences, Nashik, Maharashtra, India

Ankita Soni
Department of Pharmaceutics, Gurunanak College of Pharmacy and Technical Institute, Nagpur 440026, India

Khemchand R. Surana
Department of Pharmaceutics, Sandip Institute of Pharmaceutical Sciences, Nashik, Maharashtra, India

Swati Gokul Talele
Department of Pharmaceutics, Sandip Institute of Pharmaceutical Sciences, Nashik, India

Mohamad Taleuzzaman
Pharmaceutical Chemistry Department, and Maulana Azad University, Jodhpur 342802, India

Ana C. F. Ribeiro
Department of Chemistry, University of Coimbra, Coimbra 3004-535, Portugal

Abbreviations

AD	Alzheimer's disease
ADMET	absorption, distribution, metabolism, excretion, and toxicological
ARI	aldose reductase inhibitor
ASDs	autistic spectrum disorders
bDHC	bis-dehydroxycurcmin
bDMC	bis-demethoxycurcumin
CMC	comprehensive medicinal chemistry
Curc	monoanionic curcumin
CH	chitosan
DA	dopamine
DAT	dopamine transporter
DL	drug loading
DM	diabetes mellitus
DMA	desmethylangolensin
EA	ellagic acid
EA-NS	ellagic acid-nanosponges
eNOS	endothelial NO synthase
FTIR	Fourier transform infrared spectroscopy
GA	gallic acid
GAD	GA-enriched delivering system
GA-SiO_2	gallic acid loaded silica nanoparticles
GERD	gastroesophageal reflux disease
GI	gastrointestinal
GIT	gastrointestinal tract
GSH	glutathione
HE	hydroalcoholic extract
HHcy	homocysteine
IBS	irritable bowel syndrome
IL	interleukin
INN	International Nonproprietary Name
IR	insulin receptor
LAR	leukocyte antigen-related

LPS	lipopolysaccharide
MDDR	Modern Drug Data Report
MTD	maximum tolerated dose
NDs	neurodegenerative issue
MW	molecular weight
OS	oxidative stress
PDB	protein data bank
PDT	photodynamic therapy
PIH	post-inflammatory hyperpigmentation
QUR	quercetin
RCS	reactive carbonyl species
REOS	Rapid Elimination of Swill
RES	resveratrol
RO5	rule of 5
ROS	reactive oxygen species
SARs	structure–activity relationships
SN	substantia nigra
SP	substance-P
T2DM	type-2 diabetes mellitus
TEM	transmission electron microscopy
TPP	tripolyphosphate
USAN	United States Adapted Name
WS	*Withania somnifera*
XRD	X-ray diffraction
ZP	zeta potential

Preface

Pharmacists and nutritionists working with natural products, food scientists, nutrition researchers, and those interested in the development of innovative products, nutraceuticals, pharmaceuticals, and functional foods are sure to benefit from this thorough resource.

Pharmaceuticals are a product of scientific research that supports their claims for health improvement. Both pharmaceutical and nutraceutical compounds might be used to cure or prevent diseases, but only pharmaceutical compounds have governmental sanction.

The book emphasizes the great need for both nutritionists and pharmacologists to understand how these drugs can benefit human health.

Natural pharmaceuticals demonstrate the connections between agrochemicals and pharmaceuticals and explore the use of plants and plant products in the formulation and development of pharmaceuticals.

Functional foods are sometimes called nutraceuticals, a portmanteau of nutrition and pharmaceutical, and can include food that has been genetically modified. The use of nutraceuticals and functional foods in prevention efforts could lead to a decreased dependency on drugs. The pharmaceutical industry recognizes this shift; however, serious concerns have arisen regarding the claimed efficacy, quality, and safety of products used as medical foods.

In the first chapter, it is intended to describe that nitric oxide (NO) is a short-lived, small, highly diffusible, reactive, free radical gas, ubiquitous bioactive molecule, and derived from L-arginine (a well-known amino acid). This constituent was discovered 30 years back as "endothelium-derived relaxing factor". In mammalian cells, NO acts as a mediator and is believed to play a crucial function in several biological processes. The chapter comprehensively highlighted the emerging perspectives of natural chalcone-based nitric acid inhibitors such as sofachalcone, brussochalcone A, cardamonin, flavokawain B, dimethyl cardamonin (2′,4′-dihydroxy-6′-methoxy-3′,5′-dimethylchalcone), mallotophilippens, Hidabeni chalcone, okanin, sappanchalcone, 3-deoxysappanchalcone (3-DSC), 2′,4′,6′-tris(methoxymethoxy) chalcone (TMMC), butein, and licochalcone A which selectively inhibited the production of NO (iNOS, nNOS, eNOS), cytokine, interleukin, TNF-α, MCP-1, and prostaglandins by preventing the phosphorylated IκBα-induced translocation of NF-κB p65 subunit at nuclear milieu, inhibition of NF-κB functions, inhibiting LPS-induced translocation by Erk-1/2 MAP-kinase phosphorylation, restricting the STAT1 expression, and inducing the expression of HO-1 by activation of AKT/mTOR pathway in LPS-stimulated RAW 264.7 cells and 3T3-F442A adipocytes.

Emblicanin-A and Emblicanin-B constituents found in *Phyllanthus emblica*, also known as *Emblica officinalis* (Family: Euphorbiaceae) belonging to class hydrolysable tannins, apart from these, constituents of different classes like alkaloids, amino acids, carbohydrates, flavonoids, organic acids, phenols, and vitamins are present in the plant. Several pharmacological activities have been well-researched which include analgesic, antibacterial, larvicidal, antifungal, antioxidant, anti-inflammatory, mosquitocidal, and anticancer activities. Emblicanin-A and Emblicanin-B and as well as other has less solubility in water, that's the reason its bioavailability had reported less in formulation. Chapter 2 summarizes the pharmacological action and its applications in different nano-formulations dosages. It has abundant potential applications and is sure to be incorporated in the future into commercially available products and new uses/processes are lying in wait to be explored.

Helminth infections cause both morbidity and mortality in humans and animals by affecting the part of the body with the parasitic worms. These pathogenic worms are in general viewed under the microscope and only a few can be seen with the naked eye. Depending on the species, worms are broadly classified as flukeworms, tapeworms, roundworms, and trematodes. These worms are transmitted through ingestion of contaminated vegetables, drinking infected water, and consuming raw meat. The sign and symptoms of helminthiasis depend on the site of infection within the body includes immunological changes, malnutrition, and anemia. The major or minor inflammatory responses are observed in the skin, liver, lungs, and CNS. Different helminths can easily be identified through microscopic examination of eggs found in feces, by serological tests, and various antigen tests. The treatment strategies include prevention from multiplying worms and ultimately death of the parasite. Anthelmintic are the drugs exclusively used for the treatment of helminthiasis and other associated worm-induced diseases. Chapter 3 focuses on the pathophysiology and various herbal drugs used for treating the helminthiasis and their pharmacological activities are highlighted.

The objective of Chapter 4 is to show that various problems are produced by water retention such as hypertension, heart failure, hypervolenia, electrolyte disorder, etc. that are treated with the diuretics, which are first line antihypertensive therapy. They work on kidneys by increasing the concentration of salt and water that comes out through the urine. Further, salt will cause extra fluid to make up in the blood vessels and thereby raising the vital signs. Diuretics lower these signs by flushing the extra salt out of the body along with the accompanying fluid with it. However, after therapy with these synthetic diuretics produce numerous negative effects on the human body which certainly create problems. As a result, the general population is gradually moving toward the herbal plants and prefers the usage of natural diuretic. This chapter comprehensively describes several natural medicines, which mainly come in the form of pill, tinctures, or herbal tea that are at present employed as diuretics. A numbers of studies supported the diuretic properties of these natural medications. The chemical constituents present in these natural plant remedies play profound role in mediating the physiological interventions. The role of these hidden natural treasures as natural diuretics has wide applications in the treatment of assorted fluid retention issues owing to their safety, efficacy, and low cost price.

Psoriasis is an autoimmune mediate disease, one in every of the chronic skin disorder that has no permanent cure. It is characterized by itchiness, skin rashes, and red scalp. Different types of psoriasis are reported in medical literature. A large variety of artificial medicine agents have additionally been identified to cause skin disorder but their adverse results often challenges the benefit ratio. Therefore, the need of safer medication with simple suitableness is seldom needed. Chapter 5 aimed to explore the possible utilization, safety, and therapeutic perspectives of medicinal plant for the treating psoriasis and skin diseases which will be a potent, safe, and reliable medical aid in daily life.

The term nutraceuticals was derived from "nourishment" and "pharmaceuticals" by Stephen Defelice in 1989. For various sorts of illnesses, the nutraceuticals are an elective treatment. Nutraceutical way to deal with this sickness is a promising system particularly in certain regions it is more alluring than others. In this, we center around neuro issue like Parkinson's sickness, mental imbalance, neurodevelopmental disorders, and their belongings. In this situation, natural products (nutraceuticals) assume imperative job which is plant based. Chapter 6 basically analyzes the role of nutrients and cofactors, dietary adjustments and gut anomalies, probiotics and prebiotics, phytochemicals, and ecological factors so as to decide the condition of proof in nutraceutical-based disease management practices.

Chapter 7 has comprehensively focused on some very unknown natural chalcone compounds (kuwanon J, kuwanon R, kuwanon V, isoliquiritigenin, xanthoangelol, xanthoangelol D, xanthoangelol E, xanthoangelol F, xanthoangelol K, 4-hydroxyderricin, 5,4'-dihydroxy-6,7-furanbavachalcone, licochalcone A, licochalcone B, licochalcone C, licochalcone D, licochalcone E, echinatin, laxichalcone, broussochalcone, macdentichalcone, (2*E*)-1-(5,7-dihydroxy-2,2-dimethyl-2*H*-benzopyran-8-yl)-3-phenyl-2-propen-1-one, (2*E*)-1-(5,7-dihydroxy-2,2,6-trimethyl-2*H*-benzopyran-8-yl)-3-(4-methoxyphenyl)-2-propen-1-one, and abyssinone-VI-4-O-methyl ether) having tremendous potential to exhibit antidiabetic activity by selectively modulating the promising therapeutic target protein tyrosine phosphatase 1B (PTP-1B) that will prevent the degradation of insulin. In modern days, these natural product chalcone-based PTP-1B inhibitors are not under clinical use and they have not received any such attention in modern-day medicine as they are not explored clinically in terms of toxicological profiles to develop a suitable formulation. In the near future, it is expected that these chalcone-based PTP-1B inhibitors will open new avenues of diabetotherapeutics.

Gastrointestinal disorder is the term used to refer any condition or ailment that occurs within the gastrointestinal tract (also referred to as the GIT). Various natural drug (*Anethum graveolens* L., *Carum carvi*, *Cinnamomum tamala*, *Coriandrum sativum* L., *Foeniculum vulgare*, *Zingiber officinale* L., *Pinus roxburghii*, *Plumbago zeylanica*, *Punica granatum* L., *Saussurea lappa*, *Tamarindus indica*, and *Valeriana wallichii*) treatments are effective in lowering the signs and symptoms of GI disorders such as purposeful dyspepsia, constipation, and postoperative ileus, a painful situation that may additionally affect sufferers after a bowel surgery. Herbal medicines serve a valuable role in the management of patients with functional gastrointestinal disorders. Herbal remedy can also help with the not unusual GI circumstance irritable bowel syndrome. Many of the medication used to deal with GI issues are useless or purpose side effect. Herbal remedy is a safe, holistic alternative that normally has no prominent adverse effect. From Chapter 8, a correspondence between ethnopharmacological knowledge with modern scientific findings (antioxidant, anti-inflammatory, anti-ulcer, gastroprotective, etc.) and data of validated experiments have been presented.

For thousands of years in traditional medicines, excellent sources of pharmaceutical active ingredients are medicinal plants for the development of new drugs. Turmeric having the scientific name *Curcuma longa* belongs to the Zingiberaceae family which grows in the tropical and subtropical regions. A number of phytochemicals including curcumin, demethoxycurcumin, bisdemethoxy curcumin are present in the roots of turmeric. The polyphenolic crystalline yellowish-orange colored curcumin is the active ingredient in the herbal remedy. In China and India, the use of turmeric in traditional medicines is very common till today. The use of curcumin from turmeric as a folk remedy continues today. Chapter 9 comprehensively focuses on various reported pharmacologically active derivatives of curcumin and curcumin-based formulations.

Natural prodrugs are the chemical substance or compound obtained from living species like plants, animals, microorganisms, and marine sources. Modern-day natural products into practical applications are the small molecules modified with the aid of chemical synthesis and utilized for therapy, fortified food, the element of drug discovery, and for various healing purposes. Most of the medication for diverse sicknesses metabolizes rapidly by the process of first-pass effect which consequently results in drug inactivation and formation of harmful or poisonous metabolites in the human body. Most of the prodrugs obtained naturally from plants, animals, and marine resources undergo chemical changes to nontoxic compounds, in addition to prevention from drug inactivation. Chapter 10 highlights some of the updates on prodrugs obtained from natural products.

Hemorrhoids can be said to as one of the most prevalent ailments in modern days which have been linked with the noteworthy impact on the quality of life (QoL). Management of this disease by synthetic drugs and modern-day surgeries has not shown any impressive results and also they have limited therapeutic options. The herbal products are the gift given by Mother Nature. Chapter 11 comprehensively discusses the management of hemorrhoid disease and its associated symptoms using various natural pharmacotherapeutic options such as *Aesculus hippocastanum*, *Allium cepa*, *Bergenia ciliate*, *Bergenia ligulata*, *Bergenia stracheyi*, *Boswellia serrata*, *Boswellia carterii*, *Brassica rapa*, *Cestrum auriculatum*, *Cestrum hediundinum*, *Cissus quadrangularis*, *Commiphora mukul*, *Commiphora myrrha*, *Euphorbia prostrate*, *Ficus carica*,

Ginkgo biloba, Hamamelis virginiana, Juniperus oxycedrus, Juniperus sabina, Juniperus polycarpos, Juniperus communis, Mangifera indica, Melastoma malabathricum, Momordica charantia, Myrtus communis, Onosma Species, *Oryza sativa, Phlomis* Species, *Plantago ovata, Raphanus sativus, Ruscus aculeatus, Sesamum indicum, Syzygium cumini, Terminalia chebula, Triticum aestivum, Vaccinium myrtillus, Verbascum mucronatum, Verbascum latisepalum, Verbascum salviifolium, Verbascum lasianthum, Verbascum pterocalycinum,* and *Zingiber officinale*).

In Chapter 12, effect of light on transport of potassium thiocyanate in aqueous solutions is investigated in detail.

In Chapter 13, it is shown that drug likeness is a qualitative conception applied in drug design for how "drug-like" an element is related to factors such as bioavailability and extensively incorporated into the initial phases of lead and the drug discovery. It is projected from the molecular structure earlier the substance is at least synthesized and tested. A traditional technique to estimate drug likeness is to verify compliance of *Lipinski's Rule of Five*, which contains the amounts of hydrophilic groups, molecular weight and hydrophobicity. Methods to recognize drug-like molecules are grounded on their capability to discriminate known drugs from nondrugs in the groups of compounds by associating with one or more of the succeeding extensively available drug databases. There are different database to categorize drug-like molecules which are grounded on their capability to discriminate known drugs from nondrugs in the set of compounds and having different methods to assess the drug likeness. The conception of drug likeness has numerous applications in drug discovery.

Molecular docking is the computational tool of the molecules of complexes molded by different interactions of molecules. The aim of molecular docking in Chapter 14 is to predict the 3D structure. Molecular docking plays a significant role in the coherent design of drugs. To achieve an optimized conformation for both the protein and ligand so and relative orientation amongst protein and ligand, there are several types of docking used often such as rigid docking, flexible docking, and full flexible docking so that the free energy of the overall system is minimized. De novo drug design is a process in which the 3D structure of receptor is used to design newer molecules. It involves structural determination of the lead target complexes and lead modifications using molecular modeling tools.

CHAPTER 1

Chalcone (1,3-Diphenyl-2-Propene-1-One) Scaffold Bearing Natural Compounds as Nitric Oxide Inhibitors: Promising Antiedema Agents

DEBARSHI KAR MAHAPATRA[1*], SANJAY KUMAR BHARTI[2], and VIVEK ASATI[3]

[1]*Department of Pharmaceutical Chemistry, Dadasaheb Balpande College of Pharmacy, Nagpur 440037, India*

[2]*Institute of Pharmaceutical Sciences, Guru Ghasidas Vishwavidyalaya (A Central University), Bilaspur 495009, India*

[3]*Department of Pharmaceutical Chemistry, NRI Institute of Pharmacy, Bhopal 462021, India*

[*]*Corresponding author. E-mail: mahapatradebarshi@gmail.com*

ABSTRACT

Nitric oxide (NO) is a short-lived, small, highly diffusible, reactive, free radical gas, ubiquitous bioactive molecule, and is derived from L-arginine (a well-known amino acid). This constituent was discovered 30 years back as "endothelium-derived relaxing factor." In mammalian cells, NO acts as a mediator and is believed to play a crucial function in several biological processes. The current chapter comprehensively highlights the emerging perspectives of natural chalcone-based nitric acid inhibitors such as sofachalcone, brussochalcone A, cardamonin, flavokawain B, dimethyl cardamonin (2′,4′-dihydroxy-6′-methoxy-3′,5′-dimethylchalcone), mallo-tophilippens, Hidabeni chalcone, okanin, sappanchalcone, 3-deoxysap-panchalcone, 2′,4′,6′-tris(methoxymethoxy) chalcone, butein, and licochalcone A which selectively inhibited the production of NO (inducible NO synthase, neuronal NO synthase, and endothelial NO synthase), cytokine, interleukin, TNF-α, MCP-1, and prostaglandins by preventing the phosphorylated IκBα-induced translocation of NF-κB p65 subunit at nuclear milieu, inhibition of NF-κB functions, inhibiting LPS-induced translocation by Erk-1/2 MAP-kinase phosphorylation, restricting the STAT1 expression, and inducing the expression of heme oxygenase-1 (HO-1) by activation of AKT/mTOR pathway in LPS-stimulated RAW 264.7 cells and 3T3-F442A adipocytes.

1.1 NITRIC OXIDE

Nitric oxide (NO) is a short-lived, small, highly diffusible, reactive, free radical gas, ubiquitous bioactive molecule, and derived from L-arginine (a well-known amino acid).[1] This constituent was discovered 30 years back as "endothelium-derived relaxing factor."[2] In the NOS-dependent pathway, L-citrulline has been the point toward to be a secondary NO donor and can be converted to L-arginine.[3] It is known to play a pivotal role in the host defense mechanisms and is considered as an imperative component in various fungi, parasites, bacteria, and viruses.[4] NO diffusion occurs across the cellular membrane and no specific mechanism or process is known for its storage.[5]

In mammalian cells, NO acts as a mediator and is believed to play a crucial function in several biological processes such as apoptosis, neurotoxicity, metastasis, cardiovascular homeostasis, mitochondrial respiration, platelet functioning, vasodilatation, angiogenesis, invasion, blood flow, excitatory amino acid release, and neuronal message transmission.[6] NO has been reported to play a pivotal role in the pathophysiology of cancer (gastric, breast, head, colorectal, neck, and cervical).[7] The three isoforms: inducible NO synthase (iNOS), neuronal NO synthase (nNOS), and endothelial NO synthase (eNOS) have been seen to exhibit crucial functions in the mediation of imperative biological pathways.[8] Among these three isoforms: nNOS and eNOS isoforms are expressed constitutively (produced in larger quantities) under appropriate conditions and exclusively need calcium for the activation process while immunologic is not expressed constitutively but induces under immunologic activation (calcium-independent process).[9] Genes that express both the inducible and immunologic forms are specifically situated on various human chromosomes: viz. eNOS on 7, iNOS on 17, and nNOS on 12.[10]

It has recently observed that overexpression of this component leads to precipitation of both acute inflammation and chronic inflammation in tissues.[11] For the treatment of such conditions, few natural and synthetic inhibitors have been identified so far which acts as a potent NO production inhibitor by directly modulating the three isoforms.[12] The mediators of the immune system, the macrophages play a critical role in the defense system against various foreign agents and results in the production of NO.[13] The loss of macrophage activity is found to be associated with cellular activation after treating with endotoxins.[14] The major component of bacterial (gram-negative bacteria) cell walls, lipopolysaccharide (LPS) aggravates the proinflammatory cytokine production by swiftly activating the macrophages.[15] LPS-stimulated NO synthesis has been perceived to enhance the apoptotic process.[16] Glycine propionyl-L-carnitine, a well known dietary supplement has been recently observed clinically in amplifying the levels of NO.[17] NO enhances the Ca^{2+} channel currents by stimulating the guanylate cyclase component present in photoreceptor rod cells.[18] Inhibition of NO in LPS-stimulated cells is considered to be an attractive target in the management of chronic inflammation, platelet count, and thrombosis.[19]

1.2 CHALCONES

Chalcones are the chemical scaffold comprising two aromatic groups linked by an α, β-unsaturated carbonyl bridge (often referred to as benzylideneacetophenone or 1,3-diphenyl-2-propene-1-one) and is considered as one of the most privileged scaffolds in medicinal chemistry.[20] Kostanecki and Tambor in the 19th century fabricated chalcones for the first time in their laboratory by utilizing acetophenone and benzaldehyde.[21] They referred to these chalcone products as chromophoric products. They are considered as the precursor of flavonoid and isoflavonoid and are known as intermediate in the aurone synthesis of flavones.[22] Specific characteristics such as the presence of replaceable H-atoms, easy ways of computational studies, simple overall chemistry of the compound, and ability to modulate diverse therapeutic targets have made this scaffold attractive among young researchers.[23,24] This gifted scaffold by Mother Nature has been taken into attention owing to multifarious pharmacotherapeutic potential (either singly or in conjugation with benzothiophene, thiadiazole, thiophene, benzodiazepine, piperazine, pyridine, benzothiazepine, thiazole, pyrazine, benzoxazepine, pyrrole, isoxazole, pyrazole, pyrazoline, pyrimidine, furan, pyridazine, etc.) such as bacterial infections, protozoal infections, trypanosomiasis, sleep disorder, gout, anxiety, epilepsy, malaria, reduced immune response, hypertension, tuberculosis, diabetes, ulcer, leishmaniasis, cancer, fungal infections, reactive oxygen species, thrombosis, HIV, inflammation, metastasis, etc.[24–26]

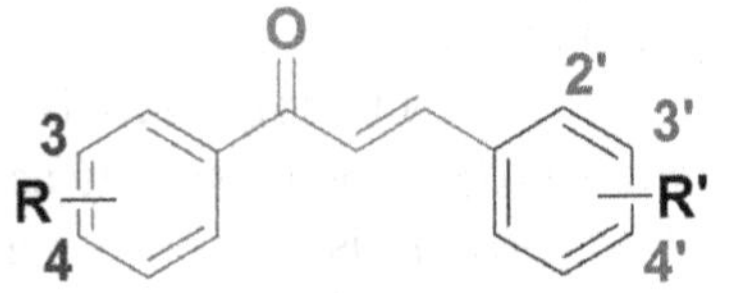

(1)

Chemical reactions play a major role in the journey of chalcones.[27] It interchanges into flavonoids in the presence of acid and the reversal into flavanone occurs by base.[28] This scaffold serves as a template structure for the elucidation of flavanones, tannins, flavonoids, and chromanochromanes.[29] Nonpharmacological utilization such as insecticides, scintillator, sweeteners in confectionaries, polymerization agents in product development, chemosensor for detection, catalyst in specific reactions, chromophores in dying industries, fluorescent polymeric agents, fluorescent whitening agent, etc. have been reported.[30] Numerous traditional high-yield methods such as Suzuki–Miyaura reaction, Claisen–Schmidt reaction, Julia–Kocienski reaction, Friedel–Crafts reaction, Sonogashira isomerization coupling, one-pot reactions, direct crossed-coupling reaction, microwave-assisted reactions, solvent-free reactions, carbonylative Heck coupling reaction, solid acid catalyst mediated reactions, etc. have been into applications in industrial-scale and academic-scale.[31,32]

1.3 CHALCONES AS NO INHIBITORS

Natural chalcone scaffold bearing compounds have been noticed as a potent inhibitor of NO and serves as anti-inflammatory agents. In addition to NO, this privileged scaffold has the attribute of modulating numerous therapeutically active targets associated directly or indirectly with inflammation.

Sofachalcone, 2'-carboxymethoxy-4,4'-bis(3-methyl-2-butenyloxy) chalcone **(1)**, has been perceived as a potent inhibitor of NO production in LPS stimulated 3T3-F442A adipocytes and RAW 264.7 macrophages culture media. In addition to the above facts, the compound has shown augmentation of HO-1 levels in the cells and also impedes the 3T3-F442A pre-adipocytes to adipocytes demarcation.[33]

Active prenyl group containing chalcone isolated from *Broussonetia papyrifera* Vent is well known as Brussochalcone A **(2)** has been found to effectually reduce the production of NO in a concentration-dependent manner in LPS-activated macrophages (IC_{50} = 11.3 μM) by suppressing the activation of NF-κB as well as iNOS expression.[34]

Chalcones obtained from *Alpinia pricei*; cardamonin **(3)**, flavokawain B (FKB) **(4)**, and dimethyl cardamonin (2′,4′-dihydroxy-6′-methoxy-3′,5′-dimethylchalcone) **(5)** and chalcones isolated from *Mallotus philippinensis* fruits; mallotophilippens (C–E) **(6–8)** have been reported to express tremendous anti-inflammatory activity by selectively inhibiting the production of NO, cytokine, interleukin, and prostaglandins by preventing the phosphorylated IκBα-induced translocation of NF-κB p65 subunit at nuclear milieu as well as inhibition of NF-κB functions.[35–38]

Chalcone glycosides isolated from *Brassica rapa* L. have been used as a template for the fabrication of 10 synthetic compounds. These chalcones popularly known as "Hidabeni" chalcones in Japanese have presented LPS-induced NO formation suppression in the rat microglia highly aggressively proliferating immortalized cells. Two molecules (**9 and 10**) expressed the highest inhibition of iNOS expression at IC_{50} values of 4.19 μM and 2.88 μM, via restricting the STAT1 expression.[39]

In suppressing the iNOS expression and associated production of NO, HO-1 plays a dominating role in activated macrophages by activating the nuclear factor-erythroid 2-related factor-2 activation. Several α,β-unsaturated carbonyl analogs have been identified as anti-inflammatory agents from genus *Bidens* (Family: Asteraceae) where okanin **(11)** has received adequate attention in inducing the expression of HO-1 in RAW 264.7 macrophages. 2-Pentene and 2-pentanone (replacing the α,β unsaturated carbonyl bridge) containing semisynthetic okanin analogs were also screened for anti-inflammatory activity; however, no prominent edema reducing effects have been recognized. The study revealed the significance of α, β unsaturated carbonyl bridge in the expression of edema reducing activity by effective NO inhibition.[40]

Polar group containing chalcones: sappanchalcone **(12)**, 3-deoxysappanchalcone (3-DSC) **(13)**, and 2′,4′,6′-tris(methoxymethoxy) chalcone (TMMC) **(14)** have been isolated from *Caesalpinia sappan* and screened in LPS-stimulated RAW 264.7 cells for exploring the anti-inflammatory potentials through NO inhibition and inducing expression of HO-1 (by activation of AKT/mTOR pathway, and not activating nuclear erythroid 2-related factor).[41–43]

The natural product butein **(15)** profoundly exerted the concentration-dependent antiedema activities by suppressing the production of NO in RAW 264.7 macrophages through inducing the expression of HO-1. The compound also displayed a decrease in NF-κB's DNA binding activity in macrophages and LPS-induced translocation by Erk-1/2 MAP-kinase phosphorylation and NF-κB inhibition.[44,45]

A well-known chalcone from *Glycyrrihza inflate*, licochalcone A **(16)** has been reported to inhibit LPS-induced NO formation. In addition to it, the natural product considerably activating NF-κB transcriptional levels and reducing the serum level of TNF-α and MCP-1 in C57BL/6 mice macrophages (Fig. 1.1).[46]

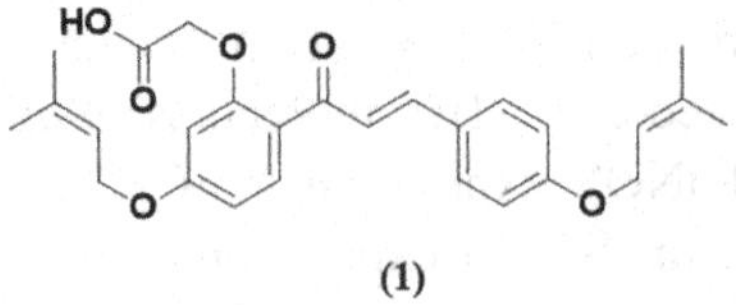

(1) 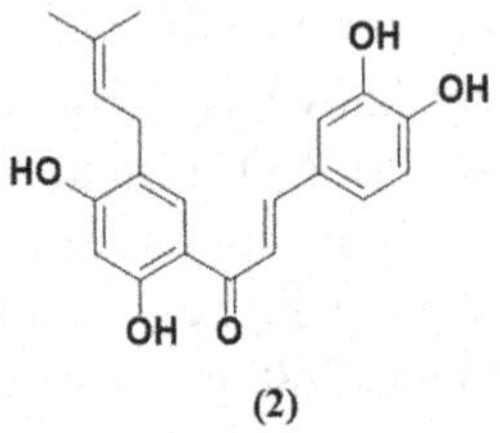 (2)

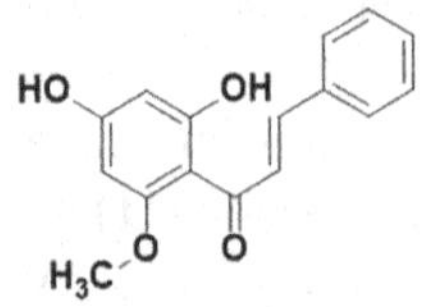

(3) 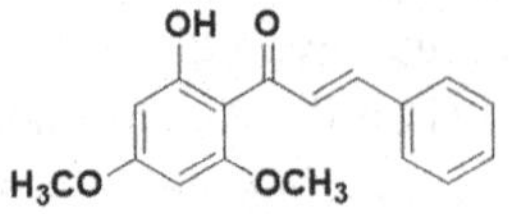 (4) 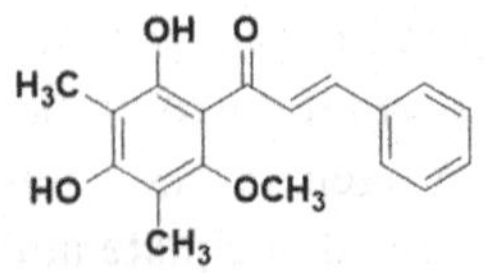 (5)

FIGURE 1.1 *(Continued)*

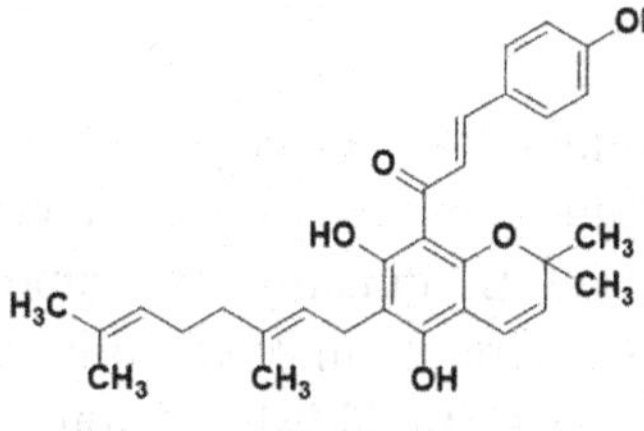

(6) 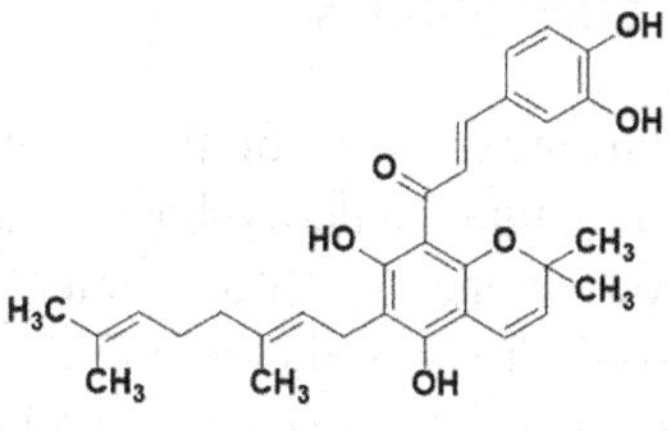 (7)

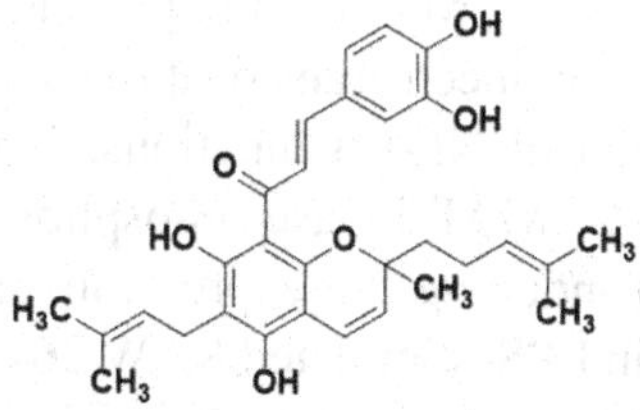

(8) 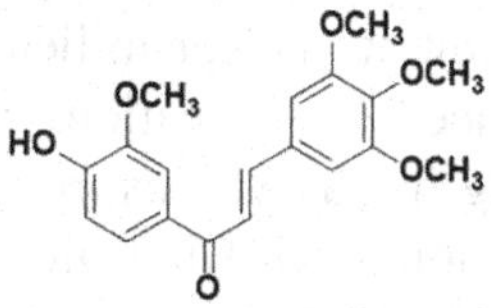 (9)

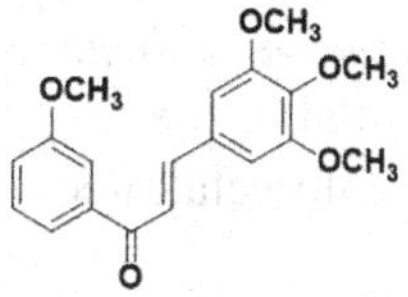

(10) 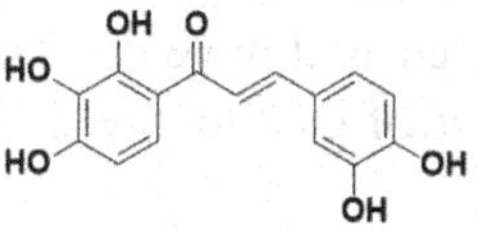 (11) 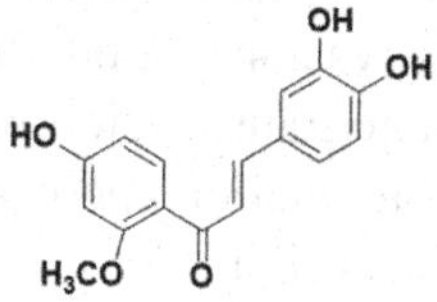 (12)

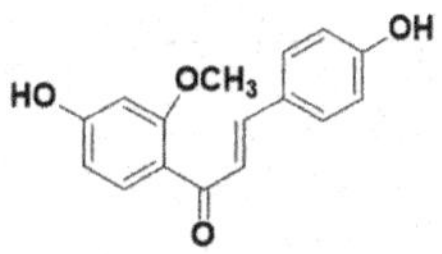

(13) 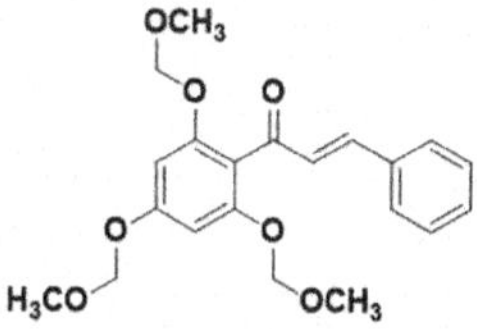 (14) 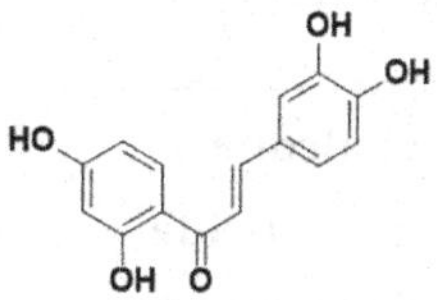 (15)

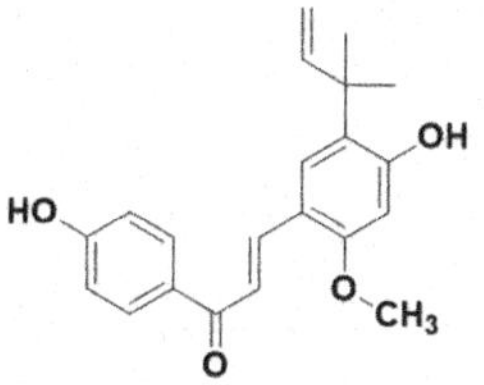

(16)

FIGURE 1.1 Structures of some naturally derived NO inhibitors as potential anti-inflammatory agents.

1.4 CONCLUSION

The current chapter comprehensively highlighted the emerging perspectives of natural chalcone-based nitric acid inhibitors such as sofachalcone, brussochalcone A, cardamonin, flavokawain B, dimethyl cardamonin (2′,4′-dihydroxy-6′-methoxy-3′,5′-dimethylchalcone), mallotophilippens, Hidabeni chalcone, okanin, sappanchalcone, 3-DSC, TMMC, butein, and licochalcone A which selectively inhibited the production of NO (iNOS, nNOS, eNOS), cytokine, interleukin, TNF-α, MCP-1, and prostaglandins by preventing the phosphorylated IκBα-induced translocation of NF-κB p65 subunit at nuclear milieu, inhibition of NF-κB functions, inhibiting LPS-induced translocation by Erk-1/2 MAP-kinase phosphorylation, restricting the STAT1 expression, and inducing the expression of HO-1 by activation of AKT/mTOR pathway in LPS-stimulated RAW 264.7 cells and 3T3-F442A adipocytes. The inhibition of NO will be helpful in modulating apoptosis, neurotoxicity, metastasis, cardiovascular homeostasis, mitochondrial respiration, platelet functioning, vasodilatation, angiogenesis, invasion, blood flow, excitatory amino acid release, and neuronal message transmission, which will have diverse potential in the treatment of inflammation and cancer of gastric, breast, head, colorectal, neck, and cervical origin.

KEYWORDS

- **chalcone**
- **nitric oxide**
- **inflammation**
- **edema**
- **natural**
- **inhibitor**
- **anti-inflammatory**
- **mechanism**

REFERENCES

1. Williams, R. J. P. Nitric Oxide in Biology: Its Role as a Ligand. *Chem. Soc. Rev.* **1996,** *25* (2), 77–83.
2. Ashutosh, K. Nitric Oxide and Asthma: A Review. *Curr. Opin. Pulm. Med.* **2000,** *6* (1), 21–25.
3. Lamattina, L.; García-Mata, C.; Graziano, M.; Pagnussat, G. Nitric Oxide: the Versatility of an Extensive Signal Molecule. *Annu. Rev. Plant Biol.* **2003,** *54* (1), 109–136.
4. Habib, S.; Ali, A. Biochemistry of Nitric Oxide. *Indian J. Clin. Biochem.* **2011,** *26* (1), 3–17.
5. Asiimwe, N.; Yeo, S. G.; Kim, M. S.; Jung, J.; Jeong, N. Y. Nitric Oxide: Exploring the Contextual Link with Alzheimer's Disease. *Oxid. Med. Cell. Longevity* **2016**.
6. Farah, C.; Michel, L. Y.; Balligand, J. L. Nitric Oxide Signalling in Cardiovascular Health and Disease. *Nat. Rev. Cardiol.* **2018,** *15* (5), 292.
7. Choudhari, S. K.; Chaudhary, M.; Bagde, S.; Gadbail, A. R.; Joshi, V. Nitric Oxide and Cancer: A Review. *World J. Surg. Oncol.* **2013,** *11* (1), 118.
8. Nathan, C.; Xie, Q. W. Nitric Oxide Synthases: Roles, Tolls, and Controls. *Cell* **1994,** *78* (6), 915–918.
9. Moncada, S. R. M. J. Nitric Oxide: Physiology, Pathophysiology, and Pharmacology. *Pharmacol. Rev.* **1991,** *43*, 109–142.
10. Wallace, J. L. Nitric Oxide as a Regulator of Inflammatory Processes. *Memorias do Instituto Oswaldo Cruz* **2005,** *100*, 5–9.
11. Sharma, J. N.; Al-Omran, A.; Parvathy, S. S. Role of Nitric Oxide in Inflammatory Diseases. *Inflammopharmacology* **2007,** *15* (6), 252–259.
12. Alderton, W. K.; Cooper, C. E.; Knowles, R. G. Nitric Oxide Synthases: Structure, Function and Inhibition. *Biochem. J.* **2001,** *357* (3), 593–615.
13. Coleman, J. W. Nitric Oxide in Immunity and Inflammation. *Int. Immunopharmacol.* **2001,** *1* (8), 1397–1406.
14. Laskin, J. D.; Heck, D. E.; Laskin, D. L. Multifunctional Role of Nitric Oxide in Inflammation. *Trends Endocrinol. Metab.* **1994,** *5* (9), 377–382.
15. Laroux, F. S.; Lefer, D. J.; Kawachi, S.; Scalia, R.; Cockrell, A. S.; Gray, L.; Heyde, H. V. D.; Hoffman, J. M.; Grisham, M. B. Role of Nitric Oxide in the Regulation of Acute and Chronic Inflammation. *Antioxid. Redox Signaling* **2000,** *2* (3), 391–396.
16. Tripathi, P.; Tripathi, P.; Kashyap, L.; Singh, V. The Role of Nitric Oxide in Inflammatory Reactions. *FEMS Immunol. Med. Microbiol.* **2007,** *51* (3), 443–452.
17. Bredt, D. S.; Snyder, S. H. Nitric Oxide: A Physiologic Messenger Molecule. *Annu. Rev. Biochem.* **1994,** *63* (1), 175–195.
18. Goldstein, I. M.; Ostwald, P.; Roth, S. Nitric Oxide: A Review of its Role in Retinal Function and Disease. *Vision Res.* **1996,** *36* (18), 2979–2994.
19. Mahapatra, D. K.; Bharti, S. K. *Handbook of Research on Medicinal Chemistry: Innovations and Methodologies*; Apple Academic Press: New Jersey, 2017.
20. Mahapatra, D. K.; Bharti, S. K.; Asati, V. Anti-Cancer Chalcones: Structural and Molecular Target Perspectives. *Eur. J. Med. Chem.* **2015,** *98*, 69–114.

21. Mahapatra, D. K.; Bharti, S. K.; Asati, V. Chalcone Scaffolds as Anti-Infective Agents: Structural and Molecular Target Perspectives. *Eur. J. Med. Chem.* **2015,** *101*, 496–524.
22. Mahapatra, D. K.; Asati, V.; Bharti, S. K. Chalcones and their Therapeutic Targets for the Management of Diabetes: Structural and Pharmacological Perspectives. *Eur. J. Med. Chem.* **2015,** *92*, 839–865.
23. Mahapatra, D. K.; Bharti, S. K. Therapeutic Potential of Chalcones as Cardiovascular Agents. *Life Sci.* **2016,** *148*, 154–172.
24. Mahapatra, D. K.; Bharti, S. K.; Asati, V. Chalcone Derivatives: Anti-Inflammatory Potential and Molecular Targets Perspectives. *Curr. Top. Med. Chem.* **2017**, *17* (28), 3146–3169.
25. Mahapatra, D. K.; Bharti, S. K.; Asati, V.; Singh, S. K. Perspectives of Medicinally Privileged Chalcone Based Metal Coordination Compounds for Biomedical Applications. *Eur. J. Med. Chem.* **2019,** *174*, 142–158.
26. Mahapatra, D. K.; Asati, V.; Bharti, S. K. An Updated Patent Review of Therapeutic Applications of Chalcone Derivatives (2014-Present). *Expert Opin. Ther. Pat.* **2019,** *29* (5), 385–406.
27. Mahapatra, D. K.; Asati, V.; Bharti, S. K. Natural and Synthetic Prop-2-Ene-1-One Scaffold Bearing Compounds as Molecular Enzymatic Targets Inhibitors against Various Filarial Species. In *Biochemistry, Biophysics, and Molecular Chemistry: Applied Research and Interactions*; Torrens, F., Mahapatra, D. K., Haghi, A. K., Eds. Apple Academic Press: New Jersey, 2019.
28. Mahapatra, D. K.; Asati, V.; Bharti, S. K. Promising Anti-Cancer Potentials of Natural Chalcones as Inhibitors of Angiogenesis. In *Natural Products Chemistry: Biomedical and Pharmaceutical Phytochemistry*; Volova, T. G., Mahapatra, D. K., Khanna, S., Haghi, A. K., Eds. Apple Academic Press: New Jersey, 2020.
29. Mahapatra, D. K.; Asati, V.; Bharti, S. K. Chalcone Scaffold bearing Natural Anti-gout Agents. In *Natural Pharmaceuticals and Green Microbial Technology: Health Promotion and Disease Prevention*; Mahapatra, D. K., Haghi, A. K., Eds. Apple Academic Press: New Jersey, 2020.
30. Mahapatra, D. K.; Bharti, S. K.; Asati, V. Recent Perspectives of Chalcone Based Molecules as Protein Tyrosine Phosphatase 1B (PTP-1B) Inhibitors. In *Medicinal Chemistry with Pharmaceutical Product Development;* Mahapatra, D. K., Bharti, S. K., Eds. Apple Academic Press: New Jersey, 2019.
31. Mahapatra, D. K.; Bharti, S. K.; Asati, V. Recent Advancements in the Pharmacotherapeutic Perspectives of Some Chalcone Scaffold Containing Natural Compounds as Potential Anti-Virals. In *Natural Products Pharmacology and Phytochemicals for Health Care: Methods and Principles in Medicinal Chemistry*; Mahapatra, D. K., Aguilar, C. N., Haghi, A. K., Eds. Apple Academic Press: New Jersey, 2020.
32. Mahapatra, D. K.; Asati, V.; Bharti, S. K. Anti-Inflammatory Perspectives of Chalcone based NF-κB Inhibitors. In *Pharmacological Perspectives of Low Molecular Weight Ligands*; Mahapatra, D. K., Bharti, S. K., Eds. Apple Academic Press: New Jersey, 2020
33. Tanaka, H.; Nakamura, S.; Onda, K.; Tazaki, T.; Hirano, T. Sofalcone, an Anti-Ulcer Chalcone Derivative, Suppresses Inflammatory Crosstalk Between Macrophages

and Adipocytes and Adipocyte Differentiation: Implication of Heme-Oxygenase-1 Induction. *Biochem. Biophys. Res. Commun.* **2009,** *381* (4), 566–571.

34. Cheng, Z. J.; Lin, C. N.; Hwang, T. L.; Teng, C. M. Broussochalcone A, A Potent Antioxidant and Effective Suppressor of Inducible Nitric Oxide Synthase in Lipopolysaccharide-Activated Macrophages. *Biochem. Pharmacol.* **2001,** *61* (8), 939–946.
35. Hatziieremia, S.; Gray, A. I.; Ferro, V. A.; Paul, A.; Plevin, R. The Effects of Cardamonin on Lipopolysaccharide-Induced Inflammatory Protein Production and MAP Kinase and NFκB Signalling Pathways in Monocytes/Macrophages. *Br. J. Pharmacol.* **2006,** *149* (2), 188–198.
36. Kim, Y. J.; Ko, H.; Park, J. S.; Han, I. H.; Amor, E. C.; Lee, J. W.; Yang, H. O. Dimethyl Cardamonin Inhibits Lipopolysaccharide-Induced Inflammatory Factors Through Blocking NF-κB p65 Activation. *Int. Immunopharmacol.* **2010**, *10* (9), 1127–1134.
37. Lin, C. T.; Senthil Kumar, K. J.; Tseng, Y. H.; Wang, Z. J.; Pan, M. Y.; Xiao, J. H.; Chien, S. J.; Wang, S. Y. Anti-Inflammatory Activity of Flavokawain B from Alpinia Pricei Hayata. *J. Agric. Food Chem.* **2009**, *57* (14), 6060–6065.
38. Daikonya, A.; Katsuki, S.; Kitanaka, S. Inhibition of Nitric Oxide Production by Novel Chalcone Derivatives from Mallotus Philippinensis (Euphorbiaceae). *Chem. Pharm. Bull.* **2004,** *52* (11), 1326–1329.
39. Hara, H.; Ikeda, R.; Ninomiya, M.; Kamiya, T.; Koketsu, M.; Adachi, T. Newly Synthesized 'Hidabeni' Chalcone Derivatives Potently Suppress LPS-Induced NO Production via Inhibition of STAT1, but Not NF-κB, JNK, and p38, Pathways in Microglia. *Biol. Pharm. Bull.* **2014,** *37* (6), 1042–1049.
40. Kil, J. S.; Son, Y.; Cheong, Y. K.; Kim, N. H.; Jeong, H. J.; Kwon, J. W.; Lee E.J.; Kwon, T. O.; Chung, H. T.; Pae, H. O. Okanin, a Chalcone Found in the Genus Bidens, and 3-Penten-2-One Inhibit Inducible Nitric Oxide Synthase Expression via Heme Oxygenase-1 Induction in RAW264.7 Macrophages Activated with Lipopolysaccharide. *J. Cin. Biochem. Nutr.* **2011,** 1112130134-1112130134.
41. Jung, E. G.; Han, K. I.; Kwon, H. J.; Patnaik, B. B.; Kim, W. J.; Hur, G. M.; Nam, K.W.; Han, M. D. Anti-Inflammatory Activity of Sappanchalcone Isolated from Caesalpinia sappan L. in a Collagen-Induced Arthritis Mouse Model. *Arch. Pharmacal Res.* **2015,** *38* (6), 973–983.
42. Kim, J. H.; Choo, Y. Y.; Tae, N.; Min, B. S.; Lee, J. H. The Anti-Inflammatory Effect of 3-Deoxysappanchalcone is Mediated by Inducing Heme Oxygenase-1 via Activating the AKT/mTOR Pathway in Murine Macrophages. *Int. Immunopharmacol.* **2014,** *22* (2), 420–426.
43. Lee, S. H.; Seo, G. S.; Kim, J. Y.; Jin, X. Y.; Kim, H. D.; Sohn, D. H. Heme oxygenase 1 mediates anti-inflammatory effects of 2′, 4′, 6′-tris (methoxymethoxy) chalcone. *Eur. J. Pharmacol.* **2006,** *532* (1–2), 178–186.
44. Sung, J.; Lee, J. Anti-Inflammatory Activity of Butein and Luteolin Through Suppression of NF κ B Activation and Induction of Heme Oxygenase-1. *J. Med. Food* **2015,** *18* (5), 557–564.
45. Lee, S. H.; Seo, G. S.; Sohn, D. H. Inhibition of Lipopolysaccharide-Induced Expression of Inducible Nitric Oxide Synthase by Butein in RAW 264.7 Cells. *Biochem. Biophys. Res. Commun.* **2004,** *323* (1), 125–132.

46. Furusawa, J. I.; Funakoshi-Tago, M.; Tago, K.; Mashino, T.; Inoue, H.; Sonoda, Y.; Kasahara, T. Licochalcone A Significantly Suppresses LPS Signaling Pathway Through the Inhibition of NF-κB p65 Phosphorylation at Serine 276. *Cell. Signal.* **2009,** *21* (5), 778–785.

CHAPTER 2

Emblicanin-A and Emblicanin-B: Pharmacological and Nano-Pharmacotherapeutic Perspective for Healthcare Applications

MOHAMAD TALEUZZAMAN[1*], DEBARSHI KAR MAHAPATRA[2], and DIPAK KUMAR GUPTA[3]

[1]*Pharmaceutical Chemistry Department, and Maulana Azad University, Jodhpur 342802, India*

[2]*Department of Pharmaceutical Chemistry, Dadasaheb Balpande College of Pharmacy, Nagpur 440037, India*

[3]*Department of Pharmaceutics, School of Pharmaceutical Education and Research, Jamia Hamdard, Hamdard Nagar, New Delhi 110062, India*

[*]*Corresponding author. E-mail: zzaman007@gmail.com*

ABSTRACT

Emblicanin-A and Emblicanin-B constituents are found in *Phyllanthus emblica*, also known as *Emblica officinalis* (Family: Euphorbiaceae) belonging to class hydrolyzable tannins. Apart from these, constituents of different classes like alkaloids, amino acids, carbohydrates, flavonoids, organic acids, phenols, and vitamins are present in the plant. Several pharmacological activities have been well-researched including analgesic, antibacterial, larvicidal, antifungal, antioxidant, anti-inflammatory, mosquitocidal, and anticancer activities. Emblicanin-A and Emblicanin-B as well as others have less solubility in water and that is the reason their bioavailability had reported less in formulation. This chapter summarizes the pharmacological action and their applications in different nanoformulation dosages. They have abundant potential applications and are sure to be incorporated in the future into commercially available products and new uses/processes are going to be explored.

2.1 INTRODUCTION

In Ayurveda from very ancient time *Phyllanthus emblica*, also known as *Emblica officinalis* (Family: Euphorbiaceae) was used as potent rasayanas. The derived medicine from this plant helps to promote health and longevity by improving defense against diseases.[1] The common names are "*Amla,*" "*Amlaki,*" "*Emblica,*" and "*Wonder Plant*" in a different region of the world. The importance of this fruit was identified before the beginning of civilization.[2] Habitat of the plant is central and southern India, Pakistan, Bangladesh, Sri Lanka, Malaysia, southern China, the Mascarene Islands, South East Asia, and Uzbekistan.[3,4]

The natural content of the fruit makes it different from others; it is believed to fight with several diseases and boost the immune system of the body. This is the richest source of vitamin-C with sufficient content of fibers, carbohydrate, and iron.[5] A commercial name Triphala combination of three fruits *E. officinalis*, *Terminalia belerica*, and *Terminalia chebula* is an excellent antioxidant.[6] The fruit has two hydrolyzable tannins, Emblicanin-A and Emblicanin-B.[7] When Emblicanin-A and Emblicanin-B get hydrolyzed, they give ellagic acid (EA), gallic acid (GA), glucose, and EA glucose. Phyllemblin is an ester found in this fruit, which potentiates the action of adrenalin, and also manifests the antiviral and antibacterial activity.[8] The fruit is highly nutritious because of the presence of alkaloids, amino acids, carbohydrates, flavonoids, organic acids, phenols, tannins, and vitamins, consumed as food (Table 9.1).[9] Mesocarp and endocarp that forms the hard stone which encages the seed are edible part of the fruit.[10]

The constituents are found in the different parts of the whole plant. The *leaves* contain chebulic acid, GA, chebulinic acid, EA, malic acid, chebulagic acid, phyllantidine alkaloids, and phyllantine. *The seeds contain of* phosphatides, fixed oil (saponification value 185, acid value 12.7, acetyl value 2.03, iodine value 139.5, sterol 2.70%, unsaponifiable matter 3.81%, and saturated fatty acid 7%), and a minute amount of essential oil. It contains linoleic acid (44%), linolenic acid (8.78%), steric acid (2.15%), oleic acid (28.40%), myristic acid (0.95%), and palmitic acid (2.99%). *The barks have* leukodelphinidin, tannin, and proanthocyanidin. *The roots* contain EA and lupeol.[11] Mankind has been using the plant originated substances from ancient times. All over the world traditional systems of medicine and its certain practice have been of importance. Presently, it is required to evaluate the selective drugs of herbal origin. It is important to know the information from traditional healers about their remedies and focus on their development in the formulation.

TABLE 9.1 Different Classes of Compounds in *Emblica officinalis*.

S. No.	Class of compounds	Constituents
1.	Hydrolysable tannins	Emblicanin A and B, punigluconin, pedunculagin, chebulinic acid (ellagitannin), chebulagic acid (benzopyran tannin), corilagin (ellagitannin), geraniin (dehydroellagitannin), ellagotannin
2.	Alkaloids	Phyllantine, phyllembein, phyllantidine
3.	Phenols	Gallic acids, methyl gallate, ellagic acid, trigallayl glucose
4.	Amino acids	Glutamic acid, proline, aspartic acid, alanine, cystine, lysine
5.	Carbohydrates	Pectin
6.	Vitamins	Ascorbic acid
7.	Flavonoids	Quercetin, kaempferol
8.	Organic acids	Citric acids

2.2 THERAPEUTIC EFFECTS OF *EMBLICANIN-A AND EMBLICANIN-B*

2.2.1 ANTIDIABETIC POTENTIALS

Globally, diabetes mellitus is increasing in number day by day and is one of the most common chronic diseases. The lifestyle revealed the current and future burden of diabetes. Multiple metabolic disorders are characterized by hyperglycemia which results from defects in insulin secretion, insulin action, or both. Type-2 diabetes mellitus (T2DM) is a major and growing health problem throughout the world. Peripheral insulin resistance and impaired insulin secretion are the main cause of T2DM. It is happening as a sequel to obesity, a sedentary life, and aging, resulting in hyperglycemia. Presently, medicine regimes for treatment of diabetes mellitus have certain drawbacks; therefore, there is a need for safer and more effective anti-diabetic drugs. Herbal formulations are preferred. The effects of aqueous extract *E. officinalis* seeds using experimental models were induced by a single intraperitoneal injection of freshly prepared streptozotocin (STZ) at 50 mg kg^{-1} dose. The assessment of blood glucose level (BGL) for fasting blood glucose and glucose tolerance test was done. The result indicates that each dose of *E. officinalis* aqueous extract reduces the BGL and improves glucose tolerance in both normal and diabetic animals. It was carried out with graded doses of 100–400 mg kg^{-1} of aqueous extract of *E. officinalis* seeds given to normal as well as sub-diabetic rats and mild-diabetic rats.[12]

The constituents belong to class flavonoids are available in optimum concentration in methanolic extracts of *P. emblica* fruit. The antihyperglycemic effect in STZ-induced diabetic rats was examined. The result had shown a best binding affinity and lead docking score to glycogen phosphorylase with QUR than with GA. A remarkable antihyperglycemic effect with different concentrations of QUR was perceived in STZ-induced diabetic rats with fewer side effects than metformin.[13]

The study had shown the effect of hydroalcoholic extract (HE) of fruits of *E. officinalis* on type-1 diabetic rats. The result of this study found a good relation of polyphenol concentration in HE of *E. officinalis* as an antidiabetic activity. The decrease in glucose levels failed to improve STZ-induced decrease in serum insulin at significant levels which concluded to have an antidiabetic activity because of the increased sensitivity of peripheral tissue to insulin or due to a direct insulin-like effect.[14]

2.2.2 ANTI-INFLAMMATORY EFFECTS AND ANTIOXIDATIVE PROPERTIES

The inflammatory process was reported to be associated with the generation of reactive oxygen species (ROS). Inflammation is a complex reaction in vascularized connective tissue, which is induced by the same exogenous and endogenous stimuli causing cell injury. The advantages of inflammation are clearing infection and wound healing; both inflammation and repair have tremendous potential to cause harm. The anti-inflammatory action of *E. officinalis* was examined in carrageenan-induced acute and cotton pellet-induced chronic inflammation in Sprague–Dawley rats. It has been performed by diminishing paw volume in acute inflammation and by decreasing cotton pellet-induced granulomas tissue lipid peroxidation, the granulomatous tissue mass, myeloperoxidase activity, and plasma extravasation in chronic inflammatory condition.[15] *E. officinalis* water extract has reported having an inhibitory effect on the synthesis and release of inflammatory mediators in rats.[16]

The fruit extract of *P. emblica* was examined for antioxidative and immunomodulatory properties. The experiment to examine the antioxidant abilities was carried out by using 2,2-diphenyl-1-picrylhydrazyl free radical scavenging, iron-reducing power, and metal chelating activity analysis which showed excellent antioxidative results. The immunomodulatory function investigates lipopolysaccharide in RAW 264.7 cells to present anti-inflammatory capacities.[17]

2.2.3 ANTIBACTERIAL ACTIVITY

The fruit extract of the plant was examined comparatively against G-positive and G-negative pathogenic bacteria. *Staphylococcus aureus* growth was attenuated by the extracts which produced a maximal zone of inhibition in the qualitative assay. The quantitative assay was performed by the well diffusion method. Antibacterial activity was examined in vitro by agar well diffusion method against *S. aureus*. The extraction of *E. officinalis* essential oil was done by using methanol, dichloromethane, hexane, and chloroform.[18]

In some autoimmune inflammatory diseases, the *E. officinalis* fruit extracts exhibited moderate inhibitors for bacterial triggers. Also, the extracts potentiated the activity of chloramphenicol and tetracycline against resistant bacterial strains.[19]

2.2.4 ANTIFUNGAL ACTIVITY

E. officinalis antifungal potential was reported against *Aspergillus*.[20] Fruit ethanol and acetone extracts showed moderate activity against *Fusarium exquisite* and *Candida albicans* when grisofulvin was used as a standard antibiotic.[21] The plant methanolic extract of *E. officinalis* did not show antifungal activity against phytopathogenic fungi *Aspergillus niger* F2723.[22]

2.2.5 ANTICANCER EFFICACY

E. officinalis exhibits its anticancer activities through inhibition of activator protein-1 and targets transcription of viral oncogenes responsible for the development of cervical cancer thus demonstrating its potential efficacy for treatment of human papillomavirus-induced cervical cancers.[23] The polyphenols from *P. emblica* can incorporate both cancer-preventative and antitumor properties. *E. officinalis* extract reduced the genotoxic effects of heavy metals and the carcinogen benzopyrene in murine models.[24]

The aqueous fractions of *E. officinalis* administered at 60–250 mg/kg prevented N-nitrosodiethylamine-induced hepatocellular carcinoma by ~80–100%. *E. Officinalis* extract was examined for chemoprevention of liver tumors induced by initiation with diethylnitrosamine followed by promotion with 2-acetylaminofluorene.[25] *P. Emblica* has potent free radical scavenging activities that might prevent ROS-induced DNA damage and oncogenesis.

2.2.6 LARVICIDAL AND MOSQUITOCIDAL ACTIVITY

The extract of *E. officinalis* exhibits larvicidal and pupicidal against the malarial vector, *Anopheles stephensi* showing 98% mortality rate at 100 ppm. A confirmed mosquitocidal property was found at 400 ppm and above also exerted 100% mortality (no hatchability).[26]

The *P. emblica* ethyl acetate leaf extracts have been found to exhibit larvicidal activity. The study concluded that the ethyl acetate extract of *P. emblica* exhibited the maximum larvicidal activity (99.6%) with LC_{50} (lethal concentration brings out 50% mortality) value of 78.89 ppm against the larvae of *Aedes egypti*.[27]

2.2.7 RADIOPROTECTIVE ACTIVITY

The mice exposed to different doses of gamma radiation and then treated with *E. officinalis* extracts showed less severity of symptoms of radiation sickness and mortality.[28] Similar radiation sickness has been reported in consecutively triphala-treated mice before irradiation when compared with the nondrug treated irradiated controls.[29]

2.3 EMBLICANIN-A AND EMBLICANIN-B NANOFORMULATION AND HEALTH CARE APPLICATION

The objective for the development of a nanosized formulation is to achieve high therapeutic efficacy with less toxicity. The herbal medicine from very past has been recognized as medicine by the physician for the patients as it has better efficacy and fewer side effects. The scientific approach for delivering the drug molecule to the target site in a sustained manner helps avoiding multiple dosing and produces less harm to the other normal cells or tissues. The novel drug delivery systems for herbal constituents reduce the repeated administration. A potential scope sees for the development of formulation with the help of nanotechnology. The nanocarriers play a vital role in the development of nanoformulation of herbal constituents.

The phytofabricated selenium nanoparticles have optimum efficacy and potential pharmacological action, that is, antioxidant and antimicrobial with biocompatibility. It was prepared from aqueous fruit extract of *E. officinalis.* It is characterized by UV–visible and Fourier transform infrared spectroscopy (FTIR) and its amorphous nature is confirmed by the X-ray diffraction (XRD) pattern and Raman spectroscopy. The highly stable nanosize has zeta potential (ZP) value negatively charged (–24.4 mV). More antioxidant activity than the standard antioxidant ascorbic acid and better antimicrobial activity in a broad range of foodborne pathogens have been shown. It has been found to be highly efficient on fungi followed by G-positive and G-negative bacteria.[30]

ZnO nanoparticles were synthesized by using *E. officinalis* leaf extract. It has been used as an antimicrobial formulation which was effective against the *S. aureus*, *S. paratyphi*, *V. cholerae*, and *E. coli*. This nanoparticle was characterized by an analytical technique like UV–Vis diffuse reflectance spectroscopy, photoluminescence measurements, XRD, FTIR, field emission scanning electron microscopy, and transmission electron microscopy (TEM). Also, it was found to possess photocatalytic activities.[31]

The EA microdispersion involved was prepared with an objective to turn the poor water solubility and low bioavailability. The content used only water and low methoxylated pectin as a food compatible excipient by applying spray drying technology which improved almost 30-times water solubility and 22% (w/w) drug loading (DL). Later used non-PAMAM (polyamidoamine) hydrophilic and amphiphilic dendrimers as nanocontainers for the preparation of two EA nanodispersions were achieved (60–70 nm) with 46 and 53% (w/w) DL, water-solubility 300–1000 times higher than that of free EA. This bioactive compound is a very good antioxidant and nontoxic suitable for food and biomedical applications.[32]

The nanoformulation of GA was prepared and characterized. The peripheral esterification with GA units, a GA-enriched delivering system (GAD) with remarkable antioxidant power and high potential against diseases from oxidative stress (OS) was achieved. GA is very effective for the disease triggered by OS. Its clinical applications are very limited because of poor gastrointestinal absorbability, pharmacokinetic drawbacks, and fast metabolism. The prepared polyester-based dendrimer for an absorbable carrier to protect and deliver GA have been fabricated. The ZP of –25 mV indicated the stability in solution with a tendency to form megamers and low polydispersity index. It has been showing to exhibit four-times higher intrinsic antioxidant power GAD as compared to the GA.[33]

Ellagic acid-nanosponges (EA-NS) used cyclodextrin and cross-linked by dimethyl carbonate is a nanoformulation which enhanced the solubilization efficiency of EA and to control its release to achieved better oral bioavailability. EA which is naturally found in various fruits has shown antioxidant, anticancer, and antimutagenic properties, the drawback of this polyphenolic compound has low oral bioavailability and it was improved by prepared a nano-formulation. The prepared EA-NS has been characterized by XRD, FTIR, and DSC studies and it elucidated a definite interaction of EA with NS. Successfully, improved its solubility and provided a controlled in vitro release for 24 h and 69.17% drug content which indicates a good DL of the prepared nanosponges. The animal studies data show the Area Under Curve (AUC) (1345.49 ng h mL^{-1}) of the EA-NS compared with (598.94 ng h mL^{-1}) for EA.[34]

The developed iron oxide magnetite coated with polyethylene glycol-gallic acid (Fe_3O_4-PEG-GA). GA possesses anticancer, antioxidant, and anti-inflammatory properties. As compared to the therapeutic effects of Fe_3O_4-PEG-GA with GA, the efficacy of both formulation of in vitro release profile percentage DL against human lung cancer cells (A-549), human breast cancer cells (MCF-7), human colon cancer cells (HT-29), and normal fibroblast cells (3T3) was assayed after incubation of 24 h, 48 h, and 72 h using (3-(4,5-dimethylthiazol-2-yl)-2,5-diphenyltetrazolium bromide) MTT assay. The result showed that nanoformulation has better efficacy than single GA.[35]

The Chitosan-Gallic acid (GA-CH) based on nanoformulation was used for the preparation of different pharmaceutical dosages. First, the nanoparticle CH was prepared and optimized. The formulation validated by Response Surface Methodology (RSM) employing the analysis of ZP and percentage encapsulation efficiency. The nanoparticles were prepared by ionotropic gelation using tripolyphosphate (TPP), at different combinations of CH concentration, CH:TPP ratio, and GA. The analysis performed confirmed the FTIR data with both hydrogen bond and ionic interactions of CH–TPP which allowed the encapsulation and the improvement of the stability of the active agent.[36]

The prepared Tween-80 coated chitosan nanoparticles (cGANP) for the nanoformulation of GA was studied on the maximum tolerated dose (MTD) using Wistar rats. A better result was found on MTD in brain-targeted nanoparticles, a profile of the drug as compared with the free drug. The MTD founded 750 mg/kg for GA and was increased to 825 mg/kg for cGANP.[37]

The phenolic constituents of *P. emblica* are not as effective as single but its nanoemulsion formulation for the topical administration showed a good result. The microemulsion technique was used for the preparation of nanoemulsion with hot high-pressure homogenization. The nanoemulsion was optimized by the ternary phase diagram. The formulation is composed of isopropyl myristate (0.6% w/w) and median particle size of 191.63 ± 4.07 nm with a narrow particle size distribution, a ZP of −10.19 ± 0.54 mV. High entrapment efficiency at 67.99 ± 0.87% and good stability at 4°C after 90 days of storage have been seen. From the nanoemulsion, the release of active drug was higher as compared to the aqueous formulation.[38]

GA control release achieved gallic acid loaded silica nanoparticles (GA-SiO_2). Aqueous solubility of GA is less and its autoxidation reduce the bioavailability, it can be enhanced by its nano-formulation. GA-SiO_2 was synthesized by a modified Stober method. GA conjugate with SiO_2 was confirmed by a different analytical technique like FTIR and XRD. The particle size was in the range of 30 nm and DL efficiency 89.39%

confirmed by scanning electron microscope and thermogravimetric analysis /differential scanning calorimetry, respectively which prolonged the antioxidant activity by its controlled release.[39]

The nanoparticles were prepared by the sonochemical method under atmospheric magnetic iron oxide (Fe^{2+} to Fe^{3+}), after that coated with CH and GA to produce a core-shell structure iron oxide-chitosan-gallic acid nanocarriers. It was characterized by analytical technique XRD and TEM. The advantage of the nanostructure had enhanced thermal stability and delivered the active drug in a controlled manner. It was nontoxic in a normal human fibroblast (3T3) line, and anticancer activity was higher in HT-29 than MCF-7 cell lines.[40]

It was well known from earlier, that many plants are rich in polyphenols with antioxidant, antiatherogenic, antidiabetic, anticancer, antiviral, and anti-inflammatory properties. It has secondary metabolites with an ability to donate electrons to free radicals through different mechanisms. The different type of nanoformulation of several constituents has been prepared and its therapeutic efficacy was observed against many diseases.[41]

The dendrimers gallic acid-triethylene glycol has been synthesized and has shown gene delivery applications, as they efficiently complex nucleic acids and form small and homogeneous dendriplexes. The interactions of the engineered dendriplexes with the blood components, as well as their stability, cytotoxicity, and the ability to enter and transfect mammalian cells have been reported. pDNA protected from degradation, and are biocompatible with HEK-293T cells and erythrocytes and effectively internalized by HEK-293T cells. The degree of PEGylation in the dendriplexes containing dendrimer/block copolymer mixtures remains an imperative parameter that requires modulation for obtaining the most optimized stealth formulation that possess the ability to effectually induce the encoded protein expression.[42]

The developed QUR and resveratrol (RES) loaded nanoformulation and *Sesbania grandiflora* leaf extract have been analyzed by a sensitive high-performance thin layer chromatography (HPTLC) method and founded 26.13 ± 0.7 μg/mg and 4.31 ± 0.8 μg/mg, respectively in leaf extract QUR and RES. The method was checked to be pH-dependent stable. The developed method was validated as per the International Council for Harmonisation of Technical Requirements for Pharmaceuticals for Human Use (ICH) guideline and it was further used in the development of nanoformulation.[43]

The natural products-based nanoformulations for the treatment of many metabolic syndromes have developed and have grown attention in the field of research. Nanosizings of the compounds enhanced the solubility, bioavailability, and promisingly enhance their efficacy. Successfully, the result of several natural constituent nanoformulation against the treatment of many diseases has been seen. The molecular targets were relevant to metabolic disorders which are affected by these compounds.[44]

QUR oil-in-water nanoemulsion developed with an objective to enhance bioavailability. It was used for the treatment of lung cancer delivered to the lung via pulmonary route. Spherical shape nanoemulsion was formed by using palm oil ester/ricinoleic acid as an oil phase, the novelty of this formulation. The nanoformulation was analyzed, exhibited high drug entrapment efficiency, and good stability against phase separation and storage at temperature 4°C for 3 months. In vitro study had shown 26.75% release in up to 48 h of QUR from nanoemulsion formulation. The cytotoxicity of nanoemulsion toward A-549 lung cancer cells without affecting the normal cells was optimized.[45]

The developed Quercetin-loaded poly(lactic-co-glycolic acid) nanoparticles formulation had shown an enhanced oral bioavailability (523% relative increase) as compared to the QUR suspension, observed in a pharmacokinetic study with a 6-day sustained release. Dosing frequency drastically decreased for the therapeutic effects.[46]

2.4 CONCLUSION

Emblicanin-A and Emblicanin-B are natural bioactive compounds that have high potential as a therapeutic agent and can be incorporated in the nano delivery systems for the treatment of several diseases. There are Emblicanin-A and Emblicanin-B and other constituents loaded nanoformulations developed were suitably evaluated for antibacterial, antifungal, analgesic, anticancer, and cerebral ischemia. The different constituents of *E. officinalis* developed nanoformulations have shown greater therapeutic effect after encapsulation process. The research on the properties of its constituent of *E. officinalis* has been continuing. Several animals and human activity have also been reported for the different class constituents of *E. officinalis*. They have abundant potential applications and are sure to be incorporated in the future into commercially available products and new uses/processes are going to be explored.

KEYWORDS

- **Emblicanin-A**
- **Emblicanin-B**
- ***Phyllanthus emblica***
- ***Emblica officinalis***
- **pharmacology**
- **nanoformulations**
- **phytochemicals**

REFERENCES

1. Udupa, K. N.; Singh, R. H. *Clinical and Experimental Studies on Rasayana Drugs and Panchakarma Therapy*. Central Council for Research in Ayurveda and Siddha: New Delhi, 1995.
2. Goyal, R. K.; Kingsly, A. R. P.; Kumar, P.; Walia, H. Physical and Mechanical Properties of Aonla Fruits. *J. Food Eng.* **2007,** *82*, 595–599.
3. Rai, N.; Tiwari L.; Sharma, R. K.; Verma, A. K. Pharmaco-botanical Profile on Emblica officinalis Gaertn. – A Pharmacopoeial Herbal Drug. *STM J.* **2012,** *1* (1), 29–41.
4. Thilaga, S.; Largia, M. J. V.; Parameswari, A.; Nair, R.R.; Ganesh, D. High Frequency Somatic Embryogenesis from Leaf Tissue of Emblica officinalis Gaertn. - A High Valued Tree for Non-timber Forest Products. *Aus. J. Crop Sci.* **2013,** *7* (10), 1480–1487.
5. Singh, E.; Sharma, S.; Pareek, A.; Dwivedi, J.; Yadav, S.; Sharma, S. Phytochemistry, Traditional Uses and Cancer Chemopreventive Activity of Amla (Phyllanthus emblica): The Sustainer. *J. App. Pharma. Sci.* **2011,** *2* (1), 176–183.
6. Phetkate, P.; Kummalue, T.; U-pratya, Y.; Kietinun, S. Significant Increase in Cytotoxic T Lymphocytes and Natural Killer Cells by Triphala: A Clinical Phase I Study. *Evid. Based Complementary Altern. Med.* **2012** , *2012*, 1–6.
7. Bhattacharya, K.; Bhattacharya, D.; Sairam, K.; Ghosal, S. Effect of Bioactive Tannoid Principles of Emblica officinalis on Ischemia-Reperfusion-Induced Oxidative Stress in Rat Heart. *Phytomedicine* **2002,** *9* (2) 171–174.
8. Yi-Fei, W.; Ya-Fenga, W.; Xiao-Yana, W.; Zhea, R.; ChuiWena, Q.; YiChenga, L. et al. Phyllaemblicin B Inhibits Coxsackie Virus B3 Induced Apoptosis and Myocarditis, *Antiviral Res.* **2009,** *84*,150–58.

9. Jain, S. K.; Khurdiya, D. S. Vitamin C Enrichment of Fruit Juice Based Ready-to-Serve Beverages Through Blending of Indian Gooseberry (*Emblica officinalis* Gaertn.) Juice. *Plant Foods Hum. Nutr.* **2004,** *59* (2):63–66.
10. Patel, S. S.; Goyal, R. K. *Emblica officinalis* Geartn. A Comprehensive Review on Phytochemistry, Pharmacology and Ethnomedicinal Uses. *Res. J. Med. Plant* **2011,** *2011*, 1–11.
11. Khan, K. H. Roles of *Emblica officinalis* in Medicine - A Review. *Bot. Res. Int.* **2009,** *2* (4), 218–228
12. Mehta, S.; Singh, R. K.; Jaiswal, D.; Rai, P. K.; Watal, G. Anti-diabetic Activity of Emblicaofficinalis in Animal Models. *Pharm. Biol.* **2009,** *47* (11), 1050–1055.
13. Srinivasan, P.; Vijayakumar, S.; Kothandaraman, S.; Palani, M. Anti-diabetic Activity of Quercetin Extracted from Phyllanthus emblica L. Fruit: In Silico and In Vivo Approaches. *J. Pharm. Anal.* **2018,** 8(2), 109–118.
14. Patel, S. S.; Goyal, R. K.; Shah, R. S.; Tirgar, P. R.; Jadav, P.D. Experimental Study on Effect of Hydroalcoholic Extract of *Emblica officinalis* Fruits on Glucose Homeostasis and Metabolic Parameters. *AYU* **2013,** *34* (4), 440–444.
15. Muthuraman, A.; Sood, S; Singla, S. K. The Antiinflammatory Potential of Phenolic Compounds from Emblica officinalis L. in Rat. *Inflammopharmacology* **2011,** *19*, 327–334.
16. Jaijoy, K.; Soonthornchareonnon, N; Panthong, A; Sireeratawong S. Anti-inflammatory and Analgesic Activities of the Water Extract from the Fruit of Phyllanthusemblica Linn. *Int. J. App. Res. Nat. Prod.* **2010,** *3* (2), 28–35.
17. Wang, H. M.; Fu, L.; Cheng, C. C.; Gao, R.; Lin, M. Y.; Su, H. L.; Belinda, N. E.; Nguyen, T. H.; Lin, W. H.; Lee, P. C.; Hsieh, L. P. Inhibition of LPS-Induced Oxidative Damages and Potential Anti-Inflammatory Effects of Phyllanthusemblica Extract via Down-Regulating NF-κB, COX-2, and iNOS in RAW 264.7 Cells. *Antioxidants* **2019,** *8*, 270.
18. Saxena, R.; Patil, P. In Vitro Antibacterial Activity of *Emblica officinalis* Essential Oil Against Staphylococcus aureus. *Int. J. Theor. Appl. Sci.* **2014,** *6* (2), 7–9.
19. Hutchings, A.; Cock, I. E. The Interactive Antimicrobial Activity of *Embelica officinalis* Gaertn. Fruit Extracts and Conventional Antibiotics Against Some Bacterial Triggers of Autoimmune Inflammatory Diseases. *Pharmacog. J.* **2018,** *10* (4), 654–662.
20. Satish, S.; Mohana, D. C.; Ranhavendra, M. P.; Raveesha, K. A. Antifungal Activity of Some Plant Extracts Against Important Seed Borne Pathogens of Aspergillus sp. *J. Agric. Technol.* **2007,** 3, 109–119.
21. Hossain, M. M.; Mazumder, K.; Hossen, S. M. M.; Tanmy, T. T.; Rashid, M. J. In *Vitro* Studies on Antibacterial and Antifungal activities of *Emblica Officinalis. IJPSR,* **2012,** *3* (4), 1124–1127.
22. Bobbarala, V.; Katikala, P. K.; Naidu, K. C.; Penumajji, S. Antifungal Activity of Selected Plants Extracts Against Phytopathogenic Fungi *Aspergillus niger*. *Indian J. Sci. Technol.* **2009,** 2(4), 87–90.
23. Mahata, S.; Pandey, A.; Shukla, S.; Tyagi, A.; Husain, S. A.; Das, B. C.; Bharti, A. C. Anticancer Activity of Phyllanthusemblica Linn. (Indian Gooseberry): Inhibition of Transcription Factor AP-1 and HPV Gene Expression in Cervical Cancer Cells. *Nutr. Cancer* **2013,** *65* (1), 88–97.

24. Jeena, K. J.; Joy, K. L.; Kuttan R. Effect of *Emblica officinalis, Phyllanthus amarus* and *Picrorrhiza kurroa* on N-nitrosodiethylamine Induced Hepatocarcinogenesis. *Cancer Lett.* **1999,** *136* (1), 11–16.
25. Sultana, S.; Ahmed, S.; Jahangir, T. *Emblica officinalis* and Hepatocarcinogenesis: A Chemopreventive Study in Wistar Rats. *J. Ethnopharmacol.* **2008,** *118* (1), 1–6.
26. Murugan, K.; Madhiyazhagan, P.; Nareshkumar, A.; Nataraj, T.; Dinesh, D.; Hwang, J. S.; Nicoletti, M. Mosquitocidal and Water Purification Properties of Ocimum Sanctum and Phyllanthus Emblica. *J. Entomol. Acarological Res.* **2012,** *44* (e17), 90–97.
27. Jeyasankar, A.; Premalatha, E. K. Larvicidal Activity of *Phyllanthus emblica Linn.* (Euphorbiaceae) Leaf Extracts Against Important Human Vector Mosquitoes (Diptera: Culicidae). *Asian Pacific J. Trop. Dis.* **2012,** *1* (2), 399–403.
28. Singh, I., Sharma, A., Jindal, A., Soyal, D., Goyal, P. K. Protective Effect of *Emblica officinalis* Fruit Extract Against Gamma Irradiation in Mice. *Pharmacologyonline* **2006,** *2*, 128–150.
29. Jagetia, G. C., Baliga, M. S., Malagi, K. J., Kamath, M. S. The Evaluation of the Radioprotective Effect of Triphala (An Ayurvedic Rejuvenating Drug) in the Mice Exposed to γ-Radiation. *Phytomedicine* **2002,** *9*, 99–108.
30. Gunti, L.; Dass, R. S.; Kalagatur, N. K. Phytofabrication of Selenium Nanoparticles From *Emblica officinalis* Fruit Extract and Exploring Its Biopotential Applications: Antioxidant, Antimicrobial, and Biocompatibility. *Front. Microbiol.* **2019,** 10, 931.
31. Maria, A.; Mookkaiahb, R.; Manikandan Elayaperumalc, M. *Emblica officinalis* Leaf Extract Mediated Synthesis of Zinc Oxide Nanoparticles for Antibacterial and Photocatalytic Activities. *Asian J. Green Chem.* **2019,** *3*, 418–431.
32. Alfei, S.; Turrini, F.; Catena, S.; Zunin, P.; Parodi, B.; Zuccari, G; Pittaluga, A. M.; Boggia, R. Preparation of Ellagic Acid Micro and Nano Formulations with Amazingly Increased Water Solubility by its Entrapment in Pectin or Non-PAMAM Dendrimers Suitable for Clinical Applications. *New J. Chem.* **2019,** *43* (6), 2438–2448.
33. Alfei, S.; Catena, S.; Turrini, F. Biodegradable and Biocompatible Spherical Dendrimer Nanoparticles with a Gallic Acid Shell and a Double-Acting Strong Antioxidant Activity as Potential Device to Fight Diseases from "Oxidative Stress". *Drug Deliv. Transl. Res.* **2019,** *10*, 259–270.
34. Mady, F. M.; Ibrahim, S. R. Cyclodextrin-based Nanosponge for Improvement of Solubility and Oral Bioavailability of Ellagic Acid. *Pak. J. Pharm. Sci.* **2018,** *31* (5), 2069–2076.
35. Rosman, R.; Saifullah, B.; Maniam, S.; Dorniani, D.; Hussein, M. Z.; Fakurazi, S. Improved Anticancer Effect of Magnetite Nanocomposite Formulation of Gallic Acid (Fe_3O_4-PEG-GA) Against Lung, Breast and Colon Cancer Cells. *Nanomaterials(Basel)* **2018,** *8* (2), 1–14.
36. Lamarra, J.; Rivero, S.; Pinotti, A. Design of Chitosan-based Nanoparticles Functionalized with Gallic Acid. *Mater. Sci. Eng. C Mater. Biol. Appl.* **2016,** *1* (67), 717–726.
37. Nagpal, K.; Singh, S. K.; Mishra, D. Influence of the Formulation on the Maximum Tolerated Doses of Brain Targeted Nanoparticles of Gallic Acid by Oral Administration in Wistar Rats. *J. Pharm Pharmacol.* **2013,** *65* (12), 1757–1764.

38. Chaiittianan, R.; Sripanidkulchai, B. Development of a Nanoemulsion of *Phyllanthus emblica L.* Branch Extracts. *Drug Dev. Ind. Pharm.* **2014,** *40* (12), 1597–1606.
39. Hu, H., Nie, L., Feng, S., Suo, J. Preparation, Characterization and In Vitro Release Study of Gallic Acid Loaded Silica Nanoparticles for Controlled Release. *Pharmazie* **2013,** *68* (6), 401–405.
40. Dorniani, D.; Hussein, M. Z.; Kura, A. U.; Fakurazi, S.; Shaari, A. H; Ahmad, Z. Preparation of Fe_3O_4 Magnetic Nanoparticles Coated with Gallic Acid for Drug Delivery. *Int. J. Nanomed.* **2012,** *7*, 5745–5756.
41. Santos, I. S.; Ponte, B. M.; Boonme, P.; Silva, A. M.; Souto, E. B. Nanoencapsulation of Polyphenols for Protective Effect Against Colon-Rectal Cancer. *Biotechnol. Adv.* **2013,** *31* (5), 514–523.
42. de la Fuente, M.; Raviña, M.; Sousa-Herves, A.; Correa, J.; Riguera, R.; Fernandez-Megia, E.; Sánchez, A.; Alonso, M. J. Exploring the Efficiency of Gallic Acid-Based Dendrimers and Their Block Copolymers with PEG as Gene Carriers. *Nanomedicine (Lond.)* **2012,** *7* (11), 1667–1681.
43. Sethuraman, V., Janakiraman, K., Krishnaswami, V., Natesan, S., Kandasamy, R. Combinatorial Analysis of Quercetin and Resveratrol by HPTLC in Sesbania Grandiflora/Phyto-Based Nanoformulations. *Nat. Prod. Res.* **2019,** *9*, 1–6.
44. Taghipour, Y. D.; Hajialyani, M.; Naseri, R.; Hesari, M.; Mohammadi, P.; Stefanucci, A.; Mollica, A.; Farzaei, M. H.; Abdollahi, M. Nanoformulations of Natural Products for Management of Metabolic Syndrome. *Int. J. Nanomed.* **2019,** *14*, 5303–5321.
45. Arbain, N. H.; Salim, N.; Masoumi, H. R. F.; Wong, T. W.; Basri, M.; Abdul Rahman, M. B. In Vitro Evaluation of the Inhalable Quercetin Loaded Nanoemulsion for Pulmonary Delivery. *Drug Deliv. Transl. Res.* **2019,** *9* (2), 497–507.
46. Chitkara, D.; Nikalaje, S. K., Mittal, A.; Chand, M.; Kumar, N. Development of Quercetin Nanoformulation and In Vivo Evaluation Using Streptozotocin Induced Diabetic Rat Model. *Drug Deliv. Transl. Res.* **2012,** *2* (2), 112–123.

CHAPTER 3

Recent In-Depth Insights of Nature-Based Anti-Worm Therapeutic Medications: Emerging Herbal Anthelmintics

ANKITA SONI[1], PARAS KOTHARI[1], and DEBARSHI KAR MAHAPATRA[2*]

[1]*Department of Pharmaceutics, Gurunanak College of Pharmacy and Technical Institute, Nagpur 440026, India*

[2]*Department of Pharmaceutical Chemistry, Dadasaheb Balpande College of Pharmacy, Nagpur 440037, India*

[*]*Corresponding author. E-mail: mahapatradebarshi@gmail.com*

ABSTRACT

Helminth infections cause both morbidity and mortality in humans and animals by affecting the parts of the body with the parasitic worms. These pathogenic worms are in general viewed under the microscope and only a few can be seen with the naked eye. Depending on the species, worms are broadly classified as flukeworms, tapeworms, roundworms, and trematodes. These worms are transmitted through ingestion of contaminated vegetables, drinking infected water, and consuming raw meat. The sign and symptoms of helminthiasis depend on the site of infection within the body includes immunological changes, malnutrition, and anemia. The major or minor inflammatory responses are observed in the skin, liver, lungs, and central nervous system. Different helminths can easily be identified through microscopic examination of eggs found in feces, by serological tests, and various antigen tests. The treatment strategies include prevention from multiplying worms and ultimately death of the parasite. Anthelmintic are the drugs exclusively used for the treatment of helminthiasis and other associated worm-induced diseases. This chapter focuses on the pathophysiology and various herbal drugs used for treating the helminthiasis and their pharmacological activities are highlighted.

3.1 INTRODUCTION

Helminth, in general the "worms" is a disease causing parasite (known as helminthiasis) that lives on or in the human or another animal. They survive by feeding on a living host by drawing nourishment from them. These worms may cause serious complications and also have the ability to multiply and invade the host immune system.[1] Helminth infections are one of the most prevalent diseases and are recognized as a major problem in several developing and developed nations.[2] The diseases caused by these helminths are chronic in nature, causes morbidity, and makes economic and social deprivation. At present, approximately 2 billion people are affected by intestinal nematodes. Depending on the species, worms are broadly classified as flukeworms, tapeworms, roundworms, and trematodes (Table 3.1). Several species of stomach worms and intestinal worms cause parasitic gastroenteritis which results in nausea, vomiting, loss of appetite, weakness, and reduced weight gain. These helminths feed themselves from their host and consume all the essential nutrients, thereby causing the retarded growth and malnutrition of the host, particularly children.[4] Other symptoms like reduced cognitive development, iron-deficiency anemia, and abdominal pains are the clinical manifesting features of severe helminth infections.[5] These worms are long lived and the mode of transmission varies with the classes of worm. It may involve ingestion of eggs or larvae, penetration by larvae, bite of vectors (dogs or cats), etc.[6] Figure 3.1 describes the morphology of *Ascaris lumbricoides*, *Trichuris trichiura*, and *Ancylostoma duodenate*.

TABLE 3.1 Classification of Worms (Helminths).

Class	Description	Examples
Nematodes	Round worms, appear round in cross section having body cavity with a straight alimentary canal and in anus	*Ascaris lumbricoides* *Trichuris trichiura* (whipworm) *Ancylostoma duodenate* (hookworm) *Enteriobius vermicularis* (threadworm)

TABLE 3.1 *(Continued)*

Class	Description	Examples
Trematodes	Leaf shaped, non-segmented, have an alimentary canal and two suckers	*Fasciolopsis* (liver fluke) *Schistosoma*
Cestodes	These are found in intestine having a head with sucking organs, they have a segmented body with no alimentary canal	*Taenia saginata* *Taenia solium* (tapeworms)

FIGURE 3.1 Pictures of *Ascaris lumbricoides*, *Trichuris trichiura*, and *Ancylostoma duodenate*.

3.2 PATHOPHYSIOLOGY OF HELMINTHS

Many infections of worm origin are asymptomatic in nature and the pathologic changes depend on the size, activity, and metabolism of the worms.

3.2.1 DIRECT DAMAGE FROM WORM ACTIVITY

The humans acquire various forms of infection by ingesting raw or uncooked meat from an infected animal host. The direct damage results from the blockage of internal organs or from the pressure exerted by the growing parasites. Tapeworms physically block the intestine and migrating *Ascaris* result in the blocking of bile duct. The eggs of these helminths can block the blood flow through many organs and leads to several pathological changes (Fig. 3.2). Exerting pressure is a general feature of larval tapeworm infections where the parasite grows as a cyst in the liver, brain, lungs, and other body cavity. These worms are having different growth form which results in metastasis and necrosis. The larvae of tapeworms develop in the CNS cavity which leads to various neurological changes and inflammation in eyes. The intestinal worms result in a variety of changes in the mucosa and tissue damage. The worms actively suck the blood that causes high blood loss from the host body.

3.2.2 INDIRECT DAMAGE FROM HOST RESPONSE

Helminths are foreign bodies, invasive, and antigenic and therefore stimulate the immune system. Adult *Schistosomes mansoni* cause infection in the blood vessels around the small intestine. These worms lay eggs in the vessels lumen and are trapped inside the liver which results in precipitation of hypersensitivity reaction by the antigens and thereby causing the formation of granuloma. The liver sinusoids become blocked which results in an impaired blood flow, leading to fibrosis of the liver. The immune-mediated changes can be clearly seen in skin, lungs, liver, and intestine. The severity of the indirect changes leads to chronic nature of infection. Another cause to infection is the soil-transmitted helminthiasis. Ascaris lumbricoides is the mostly responsible and accounts about 5–35% of all bowel obstructions. These worms are rarely noticed unless passed in stools.[8]

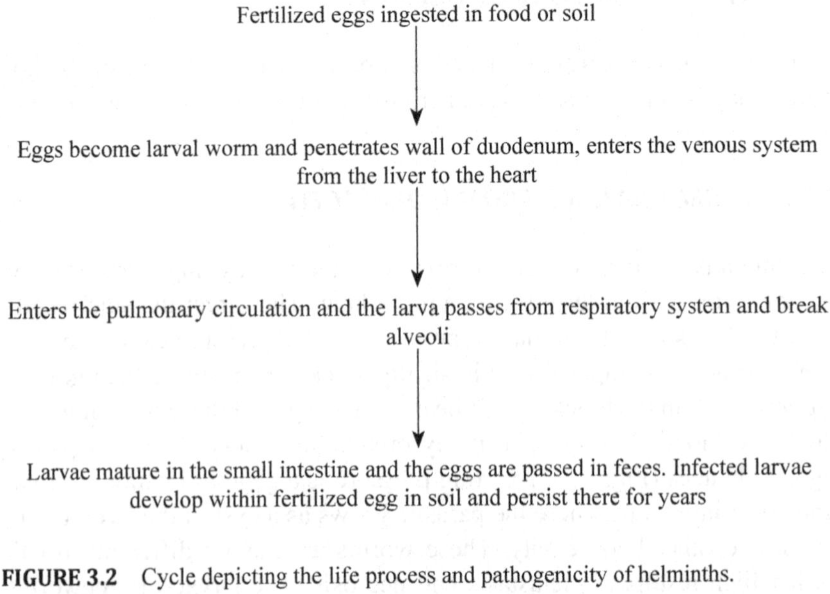

Fertilized eggs ingested in food or soil

↓

Eggs become larval worm and penetrates wall of duodenum, enters the venous system from the liver to the heart

↓

Enters the pulmonary circulation and the larva passes from respiratory system and break alveoli

↓

Larvae mature in the small intestine and the eggs are passed in feces. Infected larvae develop within fertilized egg in soil and persist there for years

FIGURE 3.2 Cycle depicting the life process and pathogenicity of helminths.

3.3 HERBAL REMEDIES USED IN ANTHELMINTICS

The herbal medicines are the most potent source of compounds with multifarious pharmacological activities. A number of plant species have shown anthelmintic activities and has attained a great interest for many researchers. The main pharmacological activity of the anthelmintic agent involves two processes that are vital to the parasites:

(1) Interference with the energy process which results in subsequent starvation of the parasite. (2) Neuromuscular blocking lead to paralysis of parasite and ultimately expulsion from the body.[4]

The phenolic compounds interfere with the energy generation process by uncoupling oxidative phosphorylation which interfere with glycoprotein of cell surface and inhibit the energy for the parasite. The tannins also bind with the free proteins present in the GIT of animals or human which provides anthelmintic activity.[9] Apart from this, alkaloids act on the CNS of the parasite and produce paralytic effect, thereby causing death of the worms. It also has antioxidant effect capable of reducing the nitrate generation and interferes in the local homeostasis, which is essential for the parasitic development.[10]

Allium sativum **(A)**, *Chenopodium album* **(B)**, *Azadirachta indica* **(C)**, *Crocus sativus* **(D)**, *Cocos nucifera* **(E)**, *Embelia ribes* **(F)**, and *Tribulus terrestris* **(G)** have been reported as emerging herbal anthelmintics (Fig. 3.3) in treating worm-based infections.

FIGURE 3.3 Reviewing imperative herbal anthelmintics.

3.3.1 ALLIUM SATIVUM (SYNONYMS: LAHASUN, LASAN, GARLIC)

It consists of dried bulbs of *Allium sativum* belonging to family Liliaceae. The bulbs (about 1.5–2.5 cm) are white to pink in color having aromatic odor and pungent taste.[11–12] Garlic contains at least 33 sulfur-containing compounds, several enzymes, 17 amino acids, and minerals. The medicinal effects and pungent odor of garlic are due to presence of sulfur compounds. The powered garlic contains 1% alliin (S-allyl cysteine sulfoxide) along with allicin (diallylthiosulfide) which is the most biologically active compound. The garlic bulb contains starch, mucilage, albumin, selenium, sugar, volatile oil, selenium, enzyme allinase, and vitamins (A, B, C, and E).[13] Garlic juice, aqueous, and alcoholic extract have tremendous broad-spectrum anthelmintic activity owing to antioxidant property, probably mediated by alliin. Apart from this, garlic has antiplatelet, antioxidant, fibrinolytic effects, and anti-atherosclerotic activities.[11]

3.3.2 CHENOPODIUM ALBUM (SYNONYMS: BATHUA, CILLISAK, VASTUK)

Chenopodium album Linn is herbaceous vegetable plant belonging to family Chenopodiaceae.[14] It is extensively found in several high plains in India and throughout Pakistan.[15] It is one of the strong smelling herbs that occur in India. It is generally 0.3–3 m high erected and has reddish, inodorous stem. The size and shape of the leaves are variable. It may sometimes be large up to 15 cm long, oblong, deltoid, obtuse, entire, toothed, or irregularly lobulate. The stems are slender, angled, striped green, red, or purple. The flowers are forming complex. The sepals are 1.5–2 mm long, oblong lanceolated. Two stigmas are present. The seeds are about 1.5 mm in diameter, compressed with an acute margin, having smooth, shining, completely annular embryo.[16]

The leaves are rich in essential oil and mineral matters, particularly in potash salts. Albuminoids, vitamin C, and nitrogen are also present. The composition includes 89.6 protein, 3.7 fat, 0.4 fiber, 0.8 carbohydrates, 2.9 minerals, 2.6 calcium, 150 phosphorous, 80 iron, 4.2 thiamine, 0.01 riboflavin, 0.14 niacin, 0.6 vitamin C, 24.0 zinc, 0.98 iodine, 6.3 fluorine, and 250 ppm vitamin K. Betalain alkaloids, phenolic acids are present in fruits, betain, and oxalic acid are found in leaves, furanocoumarins and saponins are situated in the seeds, and oleanolic acid and sitosterol are found in flowers.[17–18] The hydrodistillation of leaves yield 0.64% v/w essential oil with abundant constituents such as *p*-cymene (40.9%), ascaridole (15.5%), pinane-2-ol (9.9%), α-pinene (7.0%), β-pinene (6.2%), and α-terpinol (6.2%).[19] A new phenolic glycoside, chenoalbuside has been isolated from the methanol extract of the seeds.[20] The hydroalcoholic extract of the leaves led to the isolation of seven imperative lignans (pinoresinol, syringaresinol, lariciresinol with its derivative compound and three sesquilignans).[21] Kaempferol, quercetin, and their glycosides have been isolated from the aerial parts.[22]

The medicinal property of the plant is mainly present in leaves and seeds. The plants improve the appetite. It has remarkable anthelmintic activity against all the three classes of worms. The aqueous and ethanolic extract of leaves has wide applications in treating abdominal pain, eye disease, piles, laxative, diuretic action, throat troubles, spasmolytic, analgesic, anti-inflammatory, anticancer, hepatoprotective, diseases of heart, blood, and spleen.[16,23–26]

3.3.3 AZADIRACHTA INDICA (SYNONYMS: NEEM, NIMDO, MARGOSA TREE)

Azadirachta indica is a fast growing evergreen tree commonly found in India, Africa, and America, belonging to the family Meliaceae.[27] These are compound leaflets with oblique base, 15–25 cm long and 0.1 cm thick. The color of leaf is slightly yellowish green with bitter taste. The thickness of the bark varies according to the age having rough external surface, inner surface laminated with fracture and fibrous in nature. The flowers are arranged axillary having white color. The size of the fruit is ovoid with seed inside it having many medicinal properties.[28]

Many compounds have been isolated from the plant and have various pharmacological activities. Nimbidin, azadirachtin, nimbin, nimbolide, gedunin, mahmoodin, gallic acid, sodium nimbidinate, margolone, margolonone, isomorgolonone, cyclic trisulphide and tetrasulphide, and polysaccharides are the classic examples.[29–36] The active constituents are diterpenes (sugiol, nimbiol), triterpenes (β-sitosterol, stigmasterol, and limonoids. Flavonol glycosides are also present such as nimaton, quercetin, myrecetin, kaempferol. Neem leaves contain NLT 1.0% w/w of rutin.[37] The mature leaves contain fats, carbohydrates, proteins, amino acids, thaimine, minerals, niacine, carotene, aspartic acid.

The plant extract has been studied for gastric ulceration treatment induced by indomethacin. The antiulcer effect is due to inhibition of gastric secretion from the parietal cells. The leaf extract shows effect on serotonin inhibition in glucose mediated insulin release. The leaves are strong anthelmintic, diuretic, stomachic, astringent.[11] The plant has anti-inflammatory, antidiabetic, and antihyperlipidaemic, antibacterial, antifungal, anticarcinogenic (suppresses oral squamous cell carcinoma), antimalarial and wound-healing properties.[28,38–42]

3.3.4 *CROCUS SATIVUS (SYNONYMS: KESHAR, KUMKUM, JAFRAN, SAFFRON)*

Crocus sativus L. is a small perennial plant consist of dried stigmas and upper parts of styles, cultivated in many countries. It belongs to the family Iridaceae.[43] The color of the stigma is dark red to reddish brown and styles are yellowish brown to yellowish orange. It has a strong aromatic and characteristics odor with bitter taste. Stigmas are trifid about 25 mm long and styles are cylindrical in shape having length of 10 mm.[44] It is ethnically known to expel out worms and is a commonly recommended traditional anthelmintic for children.

More than 150 compounds have been identified with majority experimentally screened against all the three classes of worms. Colored carotenoids such as crocetin, crocins; colorless monoterpene aldehydes; safranal (responsible for odor), picrocrocin (responsible for bitter taste) are the bitter components.[45] Saffron also contains glucoside, wax, proteins, fixed oil, mucilage, sugar, β-carotene, and lycopene.[11] Crocetin is responsible for the red color of the saffron.[46]

Aqueous and ethanolic extract is investigated for reduced blood pressure activity in a dose-dependent manner. This activity is due to the effect of phytoconstituent on heart or total peripheral resistance, the later one is more important. Crocin has been investigated for the treatment of age-related macular degeneration effect.[47] Alcoholic extract of pistils of *C. sativus* L. improves learning and memory. The herbal plant has been used for its sedative, antispasmodic, expectorant, stimulant, aphrodisiac, and emmenagogue properties.[48]

3.3.5 COCOS NUCIFERA (SYNONYMS: NARIYAL, NARIKHEL, COCONUT PALM)

Cocos nucifera Linn, commonly known as coconut tree is a member of the family Arecaceae. It is largely cultivated in Southeast Asia and the island between the Indian and Pacific Oceans.[49] The plant is a monocotyledonous tree having height of about 25 m. The root is fasiculated type with unbranched stem. The leaves are feather shaped having a petiole, rachis, and leaflets. The coconut fruit is an ovoid consisting of three layers; outer epicarp, mesocarp, and inner endocarp. Inside the fruit, is solid white albumen of varied thickness with an oily pulp and a liquid known as coconut water.[50]

Coconut oil contains caprylic acid, glycerides of lauric acid, palmitic acid, and stearic acid. It also contains enzymes like oxidase, catlase, etc. The fresh kernel contains nitrogenous substances, fat, lignin, gum, albumin, tartaric acid, saccharose, myoinositol, scyllo-inositol, sorbitol, aliphatic alcohols, ketones, leucoanthocyanins, 2-propylene glycol, glycerol, glucosan, polyphenols, kampsterol, stigmasterol, and alkaloids; namely ligustrazine 2,3,5-grimethylpyrazine which play dominant role in exhibiting noteworthy anthelmintic activity.[11]

The husk fiber and the aqueous extract show outstanding analgesic activity by acting on opioid receptors. The bark extract of the tree shows antiparasitic and anti-nematocidal activity. It also has anti-inflammatory actions and is used to treat arthritis and edema. The alcoholic extract was investigated for antibacterial, antifungal, antiviral properties, and antineoplastic activity against erythroleukemia cell line.[49–51]

3.3.6 EMBELIA RIBES (SYNONYMS: BAYBIYANGA, BAVDING, BABRENG)

It consists of dried fruits of *Embelia ribes*, a shrub found in hilly regions belonging to family Myristicaeceae.[11] It is a large scandent shrub, long, slender, flexible branches with long internodes. The leaves are entire having 5–9 cm in length and 2–4 cm in height, lanceolate, short and obtusely acuminate. Numerous small flowers are present, 7.5–10 cm long with glandular pubescent. Calyx is also observed, 1.25 mm long. It has five petals having greenish yellow color, 4 mm lone, elliptic. It has five stamens, shorter than the petals and erect. The fruits are 3–4 mm in diameter, smooth, succulent, and black when ripe.[52] The fruit contains embelic acid, tannins, alkaloids, christembine, embelin, embelinol, embeliaribyl ester, embeliol, and 2,5- isobutylamine salts.[11] These compounds have been reported to express anthelmintic activity against trematodes and cestodes. Vilangin, a therapeutically active compound has also been isolated from dry ripe seeds. Nitrogen containing 3-alkyl-1,4-benzoquinone derivative, *N*-(3-carboxylpropyl)-5-amino-2-hydroxy-3-tridecyl-1,4-benzoquinone is an essential component present.[53]

3.3.7 TRIBULUS TERRESTRIS (SYNONYMS: CHOTA GOKHRU, BETAGOKHRU, LAND CALTROPS)

It consist of dried whole plant of *Tribulus terrestris*, an annual or perennial with many slender spreading branches, found throughout India, belonging to the family Zygophyceae.[11] It is small prostate, silky hairy shrub having 10–60 cm height. The leaves are unequal, pinnae from 5–8 pairs, elliptical or oblong lanceolate. The color of flowers is yellow. Its fruits are carpel, stellate, round shaped, and covered with princkles. The fresh roots are slender, fibrous, cylindrical, branched having light brown color with 7–18 cm long and 0.3–0.7 cm in diameter. The fruits are greenish yellow in color, globose, glabrous, woody cocci.[54]

The plant contains glycosides-saponins, glycosides-steroidal saponins, and steroidal sapogenins-furostenol bisglycosides, protodioscin on acid hydrolysis yields spirastanol, diosgenin, tigogenin, glucose and rhamnose, and hecogenin and nectgogenin.[55] These chemical constituents play profound role in exhibiting anti-worm activity, especially *Ascaris lumbricoides*, *Trichuris trichiura*, and *Ancylostoma duodenate*.

The presence of huge quantities of nitrates and essential oil gives the diuretic activity. The aqueous extract of the herb is useful as sexual enhancer by stimulating the testosterone production from the Leydig cells. Saponins show dose-dependent increase in phagocytosis which results in stimulation of the immune response. They also possess antidiabetic activity and decrease the postprandial blood glucose level. They are also responsible for the dilation of coronary artery and improvement in coronary circulation. β-carboline is responsible for anthelmintic, antidepressant, and anxiolytic activities.[56]

3.4 CONCLUSION

There are a number of herbal species which are responsible for producing the anthelmintic activity and in recent years their importance has been increased. Many researchers are working in this field and the good results have been established. As the demand of herbal medication is increasing day by day more advancement is accepted in coming years.

KEYWORDS

- **anthelmintic**
- **helminthiasis**
- **worm**
- **infection**
- **natural**
- **herbal**
- **traditional**

REFERENCES

1. https://www.yourgenome.org/facts/what-are-helminths
2. Krogstad, D. J.; Andengleberg, C. N. Introduction Toparasitology. Mechanisms of Microbial Disease.
3. Wen L. Y.; Yan X. L.; Sun, F. H.; Fang, Y. Y.; Yang, M. J.; Lou, L. J. A Randomized, Double-Blind, Multicenter Clinical Trial on the Efficacy of Ivermectin Against Intestinal Nematode Infections in China. *Acta Tropica*. **2008,** *106*(3), 190–194.
4. Jain, P.; Singh, S.; Singh, S. K.; Verma, S. K.; Kharya, M. D.; Solanki, S. Anthelmintic Potential of Herbal Drugs. *Int. J. Res. Dev. Pharm. Life Sci.* **2013,** *2*(3), 412–427.
5. Crompton, D. W.; Nesheim, M. C. Nutritional Impact of Intestinal Helminthiasis During the Human Life Cycle. *Annu. Rev. Nutr*. **2002,** *22*(1), 35–59.
6. Kirwan, P.; Asaolu, S.; Molloy, S.; Abiona, T.; Jackson, A.; Holland, C. Biomed Central Infectious Diseases. **2009,** *9*(20), 2334–2339.
7. https://www.ncbi.nlm.nih.gov/books/NBK8191/
8. https://www.pharmaceutical-journal.com/learning/learning-article/helminth-infections-diagnosis-and- treatment/20069529.article
9. Scalbert, A. Antimicrobial Properties of Tannins. *Phytochemistry* **1991,** *30*(12), 3875–3883.

10. Borba, H. R.; Freire, R. B.; Albuquerque, A. C.; Cardoso, M. E.; Braga, I. G.; Almeida, S. T.; Ferreira, M. J.; Fernandes, G. L.; Camacho, A. C.; Lima, R. C.; Almeida, A. C. Anthelmintic Comparative Study of Solanum lycocarpum St. Hill Extracts in Mice Naturally Infected with Aspiculuris tetraptera. *Nat. Sci.* **2010,** *8*(4), 94–100.
11. Agrawal, S. S.; Tamrakar, B. P.; Paridhavi, M. Clinically Useful Herbal Drugs; Ahuja Publishing House: Delhi, 2005; pp 163–173.
12. Kokate, C. K.; Purohit, A. P.; Gokhale, S. B.; *Pharmacognosy*, 52nd ed.; Nirali Prakashan: Pune, 2016; pp 14.45–14.46.
13. Kadam, P. U.; Kharde, M. R.; Lamage, S. T.; Borse, S. L.; Borse, L. B.; Pawar, S. P. Anthelmintic Activity of Allium Sativum. *Pharm. Sci. Monitor*. **2015,** *6*(2).
14. Kirtikar, K. R.; Basu, B. D. *Indian Medicinal Plants*, 2nd ed.; Bishen Singh Pal Singh: Delhi, 1975; pp 1465–1472.
15. Ahmad, M.; Mohiuddin, O. A.; Jahan, N.; Anwar, M. U.; Habib, S.; Alam, S. M.; Baig, I. A. Evaluation of Spasmolytic and Analgesic Activity of Ethanolic Extract of Chenopodium album Linn and its Fractions. *J. Med. Plants Res.* **2012,** *6*(31), 4691–4697.
16. Kokate, C. K. *Practical Pharmacognosy*, Vol. I, 3rd ed.; Vallabh Prakashan: New Delhi, 1994; pp 115–27.
17. Lavaud, C.; Voutquenne, L.; Bal, P.; Pouny, I. Saponins from Chenopodium album. *Fitoterapia* **2000,** *71*(3), 338–340.
18. Nicholas, H. J.; Wadkins, C. L.; Hiltibran, R. C. The Distribution of Triterpenes in Plants. Chenopodium album. *J. Am. Chem. Soc.* **1955,** *77*(2), 495–496.
19. Guil, J. L.; Torija, M. E.; Giménez, J. J; Rodríguez-García, I.; Giménez, A. Oxalic Acid and Calcium Determination in Wild Edible Plants. *J. Agric. Food Chem.* **1996,** *44*(7), 1821–1823.
20. Nahar, L.; Sarker, S. D. Chenoalbuside: An Antioxidant Phenolic Glycoside from the Seeds of Chenopodium album L.(Chenopodiaceae). *Revista Brasileira de Farmacognosia.* **2005,** *15*(4), 279–282.
21. Cutillo, F.; DellaGreca, M.; Gionti, M.; Previtera, L.; Zarrelli, A. Phenols and Lignans from Chenopodium album. Phytochemical Analysis. *Int. J. Plant Chem. Biochem. Tech.* **2006,** *17*(5), 344–349.
22. Bylka, W.; Kowalewski, Z. Flawonoidy w Chenopodium album L. i Chenopodium opulifolium L.(Chenopodiaceae). *Herba Polonica.* **1997,** *3*(43), 208–213.
23. Ankita, J.; Chauhan, R. S. Evaluation of Anticancer Activity of Chinopodium album Leaves in BHK-21 Cells. *Int. J. Univers. Pharm. Bio Sci.* **2012,** *1*(2), 92–102.
24. Vijay, N.; Padmaa, M. Hepatoprotective Activity of Chenopodium album Linn. Against Paracetamol Induced Liver Damage. *Pharmacologyonline* **2011,** *3*, 312–328.
25. Nigam, V.; Paarakh, P. M. Anti-Ulcer Effect of Chenopodium album Linn. Against Gastric Ulcers in Rats. *Int. J. Pharm. Sci. Drug Res.* **2011,** *3*(4), 319–322.
26. Usman, L. A.; Hamid, A. A.; Muhammad, N. O.; Olawore, N. O.; Edewor, T. I.; Saliu, B. K. Chemical Constituents and Anti-Inflammatory Activity of Leaf Essential Oil of Nigerian Grown Chenopodium album L. *EXCLI J.* **2010,** *9*, 181.
27. Pingale Shirish, S. Hepatoprotection Study of Leaves Powder of Azadirachta indica A. Juss. *Int. J. Pharm. Sci. Rev. Res.* **2010,** *3*(2), 37–42.
28. Maithani, A.; Parcha, V.; Pant, G.; Dhulia, I.; Kumar, D. Azadirachta indica (Neem) Leaf: A Review. *J. Pharm. Res.* **2011,** *4*(6), 1824–1827.

29. Pillai, N. R.; Santhakumari, G. Anti-Arthritic and Anti-Inflammatory Actions of Nimbidin. *Planta Medica* **1981,** *43*(09), 59–63.
30. Jones, I. W.; Denholm, A. A.; Ley, S. V.; Lovell, H.; Wood, A.; Sinden, R. E. Sexual Development of Malaria Parasites is Inhibited in vitro by the Neem Extract Azadirachtin, and its Semi-Synthetic Analogues. *FEMS Microbiol. Lett.* **1994,** *120*(3), 267–273.
31. Sharma, V. N.; Saksena, K. P. 'Sodium-Nimbidinate'--in vitro Study of its Spermicidal Action. *Indian J. Med. Sci.* **1959,** *13*, 1038.
32. Rojanapo, W.; Suwanno, S.; Somaree, R.; Glinsukon, T.; Thebtaranonth, Y. Screening of Antioxidants from Some Thia Vegetables and Herbs. *J. Sci. Thailand.* **1985,** *11*, 177–188.
33. Khalid, S. A.; Duddeck, H.; Gonzalez-Sierra, M. Isolation and Characterization of an Antimalarial Agent of the Neem Tree Azadirachta indica. *J. Nat. Prod.* **1989,** *52*(5), 922–927.
34. Ara, I.; Siddiqui, B. S.; Faizi, S.; Siddiqui, S. Structurally Novel Diterpenoid Constituents from the Stem Bark of Azadirachta indica (Meliaceae). *J. Chem. Soc. 1989, 2*, 343–345.
35. Pant, N.; Garg, H. S.; Madhusudanan, K. P.; Bhakuni, D. S. Sulfurous Compounds from Azadirachta indica Leaves. *Fitoterapia* **1986**.
36. Kakai, T.; Koho, J. P. Anti-Inflammatory Polysaccharides from Melia azadirachta. *In Chem. Abstr.* **1984,** *100*, 91350.
37. Kokate, C. K.; Purohit, A. P.; Gokhale, S. B. Pharmacognosy, 52nd ed.; Nirali Prakashan: Pune, 2016; pp 19.2–19.4.
38. Chattopadhyay, R. R. Possible Biochemical Mode of Anti-Inflammatory Action of Azadirachta indica A. Juss. in Rats. *Indian J. Exp. Biol.* **1998,** *36*(4), 418–420.
39. Bopanna, K. N.; Kannan, J.; Sushma, G.; Balaraman, R.; Rathod, S. P. Antidiabetic and Antihyperlipaemic Effects of Neem Seed Kernel Powder on Alloxan Diabetic Rabbits. *Indian J. Pharmacol.* **1997,** *1*, *29*(3), 162.
40. Thakurta, P.; Bhowmik, P.; Mukherjee, S.; Hajra, T. K.; Patra, A.; Bag, P. K. Antibacterial, Antisecretory and Antihemorrhagic Activity of Azadirachta indica Used to Treat Cholera and Diarrhea in India. *J. Ethnopharmacol.* **2007,** *22*, *111*(3), 607–612.
41. Venugopal, V. Antidermatophytic Activity of Neem (Azadirachta indica) Leaves in vitro. *Indian J. Pharmacol.* **1994,** *26*(2), 141.
42. Balasenthil, S.; Arivazhagan, S.; Ramachandran, C. R.; Ramachandran, V.; Nagini, S. Chemopreventive Potential of Neem (Azadirachta indica) on 7, 12-Dimethylbenz [a] Anthracene (DMBA) Induced Hamster Buccal Pouch Carcinogenesis. *J. Ethnopharmacol.* **1999,** *67*(2), 189–195.
43. Abdullaev, F. I. Biological Effects of Saffron. *Bio Factors* **1993,** *4*(2), 83–86.
44. Kokate, C. K.; Purohit, A. P.; Gokhale, S. B. Pharmacognosy, 52nd ed.; Nirali Prakashan: Pune, 2016; pp 14.105–14.107
45. Bathaie, S. Z.; Mousavi, S. Z. New Applications and Mechanisms of Action of Saffron and its Important Ingredients. *Crit. Rev. Food Sci. Nutr.* **2010,** *50*(8), 761–786.
46. Martin, G.; Goh, E.; Neff, A. W. Evaluation of the Developmental Toxicity of Crocetin on Xenopus. *Food Chem. Toxicol.* **2002,** *40*(7), 959–964.

47. Srivastava, R.; Ahmed, H.; Dixit, R. K. Crocus sativus L.: A Comprehensive Review. *Pharmacogn. Rev.* **2010,** *4*(8), 200.
48. Nemati, H.; Boskabady, M. H.; Vostakolaei, H. A. Stimulatory Effect of Crocus sativus (saffron) on β2- Adrenoceptors of Guinea Pig Tracheal Chains. *Phytomedicine* **2008,** *15*(12), 1038–1045.
49. Lima, E. B; Sousa, C. N.; Meneses, L. N.; Ximenes, N. C.; Júnior, S, Vasconcelos, G. S.; Lima, N. B.; Patrocínio, M. C.; Macedo, D.; Vasconcelos, S. M. Cocos nucifera (L.)(Arecaceae): A Phytochemical and Pharmacological Review. *Braz. J. Med. Biol. Res.* **2015,** *48*(11), 953–964.
50. Passos, E. E. Morfologia do coqueiro. A cultura do coqueiro no Brasil. **1998,** *2*, 57–64.
51. DebMandal, M.; Mandal, S. Coconut (Cocos nucifera L.: Arecaceae): In Health Promotion and Disease Prevention. *Asian Pac. J. Trop. Med.* **2011,** *4*(3), 241–247.
52. Kirtikar, K. R.; Basu, B. D. *Indian Medicinal Plants*, Vol. 3.
53. Indrayan, A. K.; Sharma, S.; Durgapal, D.; Kumar, N.; Kumar, M. Determination of Nutritive Value and Analysis of Mineral Elements for Some Medicinally Valued Plants from Uttaranchal. *Curr. Sci.* **2005,** *10*, 1252–1255.
54. Chhatre, S.; Nesari, T.; Somani, G.; Kanchan, D.; Sathaye, S. Phytopharmacological Overview of Tribulus terrestris. *Pharmacogn. Rev.* **2014,** *8*(15), 45.
55. Usman, H.; Abdulrahman, F. I.; Ladan, A. H. Phytochemical and Antimicrobial Evaluation of Tribulus terrestris L.(Zygophylaceae). Growing in Nigeria. *Res. J. Bio. Sci.* **2007,** *2*(3), 244–247.
56. Chhatre, S.; Nesari, T.; Somani, G.; Kanchan, D.; Sathaye, S. Phytopharmacological Overview of Tribulus Terrestris. *Pharmacogn. Rev.* **2014,** *8*(15), 45.

CHAPTER 4

Insights into the Recent Scientific Evidences of Natural Therapeutic Treasures as Diuretic Agents

VAIBHAV SHENDE[1], SAMEER HEDAOO[1], and
DEBARSHI KAR MAHAPATRA[2*]

[1]*Department of Pharmaceutics, Gurunanak College of Pharmacy and Technical Institute, Nagpur 440026, India*

[2]*Department of Pharmaceutical Chemistry, Dadasaheb Balpande College of Pharmacy, Nagpur 440037, India*

[*]*Corresponding author. E-mail: mahapatradebarshi@gmail.com*

ABSTRACT

Various problems are caused by water retention such as hypertension, heart failure, hypervolenia, electrolyte disorder, etc. that are treated with the diuretics, which are first line antihypertensive therapy. They work on kidneys by increasing the concentration of salt and water that comes out through the urine. Further, salt will cause extra fluid to make up in the blood vessels and thereby raising the vital signs. Diuretics lower these signs by flushing the extra salt out the body along with the accompanying fluid with it. However, after therapy with these synthetic diuretics produce numerous negative effects on the human body which certainly create problems. As a result, the general population is gradually moving toward the herbal plants and prefers the usage of natural diuretic. This chapter comprehensively describes several natural medicines, which mainly come in the form of pill, tinctures, or herbal tea, that are at present employed as diuretics. A numbers of studies supported the diuretic properties of these natural medications. The chemical constituents present in these natural plant remedies play profound role in mediating the physiological interventions. The pharmacological perspectives of these hidden natural treasures in treating the assorted fluid retention issues have been reported. These natural diuretics have wide applications and are preferred by individuals owing to their safety, efficacy, and low cost price.

4.1 INTRODUCTION

Diuretics are a useful and heterogeneous category of agents, most typically employed in the treatment of cardiovascular disease, heart condition, and solution disorders.[1] Diuretics alter the volume and/or composition of body fluids during clinical setting, together with high blood pressure, coronary failure, kidney disease, syndrome, and cirrhosis of the liver.[2] They act by reducing the dual compound biological process at different sites within the tubule, thereby increasing the urinary sodium, and consequently the water loss.[3] From 1919 to 1960, the foremost effective diuretics used in the mainstay of treatment were the mercurials; however, they lost their charm in the due course owing to their toxicity.[4] Alternative choices throughout that era were limited to diffusion diuretics like carbamide, osmitrol, and saccharose; acidifying salts, organic compound derivatives, and lanoxin that has a water pill result additionally to its inotropic effect.[5]

There are many artificial drug medicines that are universally used for the treatment of high blood pressure, heart condition, and different water retention disorders.[6] Modern day diuretics include ethacrynic acid, torsemide, spironolactone, hydrochlorothiazide, acetazolamide, methazolamide, amiloride, triamterene, and mannitol (Table 4.1).[7] But these medication have varied adverse effects like hypokalemia, hypomagnesemia, dilutional hyponatremia, allergic manifestations, hyperuricemia, symptom, drowsiness, fatigue, abdominal discomfort, rise in blood carbamide, nausea, dizziness, muscle cramps, headache, chills, polydipsia, confusion, and pain in chest.[8] Because of these adverse effects, the local populations have rapidly moved toward the natural therapy.[9] In the race to find a cure for water retention-related problems, many individuals are turning to natural diuretics.[10] Herbal and natural merchandise of folks medicines are used for hundreds of years in each culture throughout world.[11] Scientists and medical professionals have shown inflated interest within the field as they recognized truth health edges of those remedies.[12] Herbal medication have gained importance and recognition in recent years as a result of their safety, economicity, and value effectiveness.[13]

TABLE 4.1 Classification of Modern Day Diuretics.

Diuretic	Example	Uses	Adverse effects	Contraindication
Loop diuretics	Ethacrynic acid, torsemide	Edema, acute pulmonary edema, cerebral edema, hypertension	Hypokalemia, dilutional hyponatremia, hyperglycemia	It may enhance digitalis toxicity and can cause cardiac irregularities due to hypokalemia. With cotrimoxazole, it increases the chances of thrombocytopenia
Thiazide diuretics	Hydrochlorothiazide	Hypertension, diabetes insipidus	Hypocalcaemia, magnesium depletion	Enhances digitalis toxicity, reduces sulfonylurea action
Carbonic anhydrous inhibitors	Acetazolamide, methazolamide	Glaucoma, to alkalinize urine	Drowsiness, fatigue	Decrease the reabsorption of some acidic drugs
Potassium-sparing diuretics:		To treat edematous conditions including liver cirrhosis	Hyperkalemia, rise in blood urea, nausea, dizziness, and muscle cramps	NSAIDs can decrease the effect of these agents, quinidine with amiloride might increase the danger of arrhythmias
Aldosterone antagonists	Spironolactone			
Epithelial sodium channel blockers	Amiloride, triamterene			
Osmotic diuretics	Mannitol	Barbiturate poisoning, threatened acute renal failure	Headache, nausea, chills, polydipsia, confusion, and pain in chest	Glycerin is not utilized in diabetes, carbamide and diuretic are not utilized in intracranial hemorrhage

4.2 MECHANISM OF DIURETICS

The diuretics play a crucial role with the management of dropsy and high blood pressure.[13] This process operates the rise in negative water and substance balance.[14] The proximal convoluted tube-shaped structure reabsorbs the fluid by both active and passive processes.[15] The skinny descending loop of Henle permits the diffusion water abstraction is impermeable to the solutes and play pivotal role in reducing water absorption from the descendent limb in overall increased symptomatic condition (Fig. 4.1).[16] The skinny ascending limb of loop of Henle is rubberized to water and extremely permeable to chloride and sodium diuretics which show no pronounced effects on that.[17]

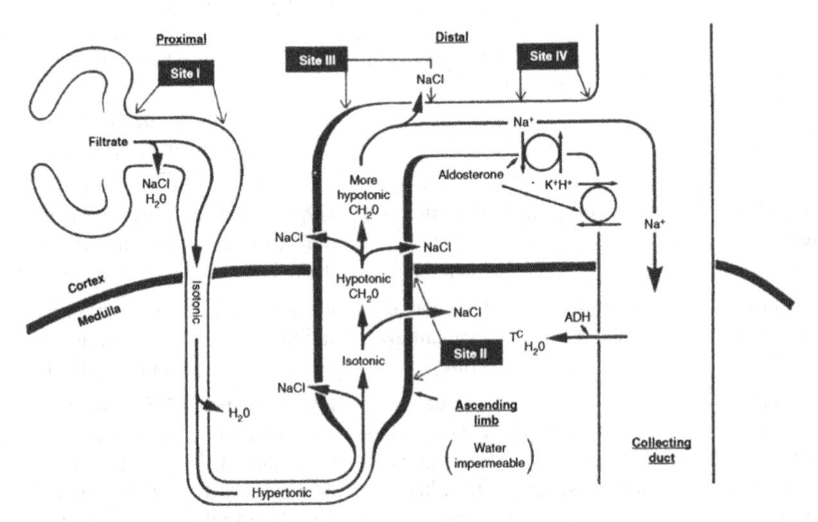

FIGURE 4.1 Physiology of urine formation and mechanism of diuretics.

4.3 ROLE OF KIDNEY IN WATER HOMEOSTASIS

The whole organic processes of filtering the fluid happen by diffusion process.[18] Since the produced capillary vessel filtrate is actually isoosmotic, it largely depends on the metal organic process to form the associate diffusion gradient.[19] Once the formation of a plasma ultrafiltrate within the capillary occur, the fluid enters the proximal convoluted tube-shaped structure, where specific transporters absorb metal, chloride, hydrogen carbonate, aldohexose, and amino acids.[20] Concerning the amount of water and most of the organic solutes, they are again reabsorbed within the proximal tube-shaped structure.[21] At the boundary between the inner and outer stripes of the outer medulla, the skinny dropping limb of Henle activity reabsorbs both sodium and chlorine from the lumen (about 3/5th of the filtered sodium).[22] However, the conditions are not like the proximal tube and therefore the dropping limb is nearly impermeable to the water.[23] The sodium chloride resorption occur within the thick ascending limb effectively, which dilutes the fluid, therefore this section is termed the "diluting segment."[24] The loop of Henle also acts in producing a countercurrent regarding the formation of a gradient of hyperosmolarity within the medullary interstitium.[25] The distal convoluted tube-shaped structure connects the diluting section, where around 100% of the filtered common salt is reabsorbed.[26] Just like the thick ascending limb, the membrane is comparatively tight to water, therefore ensues more dilution.[27]

The final step of urine composition is the grouping duct, where 2–5% of the sodium chloride reabsorption process happens significantly.[28] The mineralocorticoids exert their influence here and all diuretic-induced changes occur.[29] The water is reabsorbed through the action of the posterior pituitary gland hormone vasopressin (also referred to as vasoconstrictor [ADH], though ADH is that the most popular term) and thus the final urine entering the cavity is diluted or concentrated.[30] This is achieved by the countercurrent mechanism that makes the gradient from 50 mm/kg at the outer cortex to 1200 mm/kg at the inner medulla.[31]

4.4 USE OF HERBAL DRUGS AS DIURETICS

4.4.1 ABUTILON INDICUM

The seed extract of dilleniid dicot *A. indicum* (200 mg/kg and 400 mg/kg) were evaluated for its diuretic activity. The extract at above doses produced vital dose-dependent augmentation in urinary excretion and urinary metallic element loss.[32]

4.4.2 ACACIA SUMA

The compound extract of *A. suma* barks considerably enhances the urinary output in the concentration of 400 mg/kg, p.o.; however, the result was found to be the less potent in elevating the urinary output in comparison with the reference standard furosemide. Further, the extract was found to enhance the urinary solution concentration (Na^+, K^+, and Cl^-). The rise within the quantitative relation of concentration of excreted K^+ ions indicates that the extract increase sodium ion excretion to a larger than K^+. It may be a terribly necessary step toward a perfect water pill with lesser hyperkalemic result.[33]

4.4.3 *ACHYRANTHES ASPERA*

The results obtained during this study offer a quantitative basis to clarify the normal folkloric use of *A. aspera* as a diuretic drug agent. It is additionally used for the treatment of high blood pressure and excretory organ diseases. The extract has diuretic drug impact supporting the ethnopharmacological use as diuretics. This impact is also explored within the use of the plant in the management of some *vata* (Ayurvedic) disease.[34]

4.4.4 *AZIMA TETRACANTHA*

The study on methanolic extract of *A. tetracantha* leaf was carried out in unusual person rats to determine the diuretic activity. The extract was found to exhibit dose-dependent diuretic activity as shown by an increase in the urine output with an increase in the concentration of extract.[35]

4.4.5 *BOERHAAVIA DIFFUSA*

Researchers studied the impact of *B. diffusa* extract on urinary organ and found that the content stays are in agreement with its diuretic drug effect. The diuretic drug activity was attributed to glucosides, ecdysone, and several elements. However, the extreme diuretic activity was ascertained if the roots are taken and find application in season cares (Table 4.2). Diuretic are the primary line of medical care in nephritic inflammatory disorders because it nearly always involves derangement in fluid dynamics of the body. During this regard hepatoprotective property of the plant has another profit.[36]

4.4.6 *TARAXACUM OFFICINALE*

The sesquiterpenes lactones present in *T. officinale* are primarily responsible for diuretic drug effects and also contribute to dandelion's gentle medicine activity.[37]

4.4.7 *MACROTYLOMA UNIFLORUM*

M. uniflorum is pungent after digestion, act as astringent, strong diuretic, the tonic cures disorders related to *pitta* (Ayurvedic) and *rakta* (Ayurvedic), hot in potency, migrates dyspnea, cough perspiration, reduces fat, and cures fever.[38]

4.4.8 *CITRULLUS LANATUS*

C. lanatus is diuretic, quite hard for digestion, unctuous, very sweet, cold in potency, aphrodisiac, mitigates *pitta* (Ayurvedic) and *vata* (Ayurvedic), produces bleeding disease and dysuria especially. Fruit juice and pulp are cooling, nutritive, demulcent, and a ethnopharmacologically privileged diuretic. The root is purgative.[39]

4.4.9 *SETANIA INDICA*

S. italica unites the broken tones, aggravates *vata* (Ayurvedic), strengthens the body, is hard to digest, causes dryness, and mitigates *kapha* (Ayurvedic) greatly. It is a potent diuretic and an astringent.[40]

4.4.10 *CAMELLIA SINENSIS*

Every time when a person relishes a hot cup of normal tea or black tea, it starts flushing excess of fluid from the system and acts perfectly as a natural diuretic.[41]

4.4.11 *CORIANDER SATIVUM*

The urinary excretion of chloride ion gets enhanced by the plant extract of *C. sativum*.[42]

TABLE 4.2 Emerging Herbal Therapy for Diuretic Application.

Herbal drugs	Synonym	Biological source (family)	Parts of plant use	Traditional diuretic and other uses	Images of herbal plant
Abutilon indicum	Indian Mallow, Sida Indica L.	*Abutilon indicum* Linn (Malvaceae)	Leaves, root, flower	Diuretic and natriuretic activities, urethritis, antioxidant, antiulcer	
Acacia	Roxb Var	*Acacia polyacantha* (Fabaceae)	Barks	Diuretic, laxative	
Achyranthes Aspera	Achyranthes Aspera L.	*Achyranthes aspera* (Amaranthaceae)	Whole plant	Diuretic, spermicidal, anti-allergic, nephroprotective	

TABLE 4.2 *(Continued)*

Herbal drugs	Synonym	Biological source (family)	Parts of plant use	Traditional diuretic and other uses	Images of herbal plant
Azima tetracantha	Monetia Barlerioides	*Azima tetracantha* Linn (Salvadoraceae)	Leaf	Diuretic	
Boerhaavia	Spreading hogweed, punarnava	*B. diffusa* (Nyctaginaceae)	Leaves, stem, roots	Diuretic, renal activity	
Dandelion	Blowball, meadow herb	*Taraxacum officinale* Weber (Asteraceae)	Whole plant	Diuretic, choleretic, antirheumatic	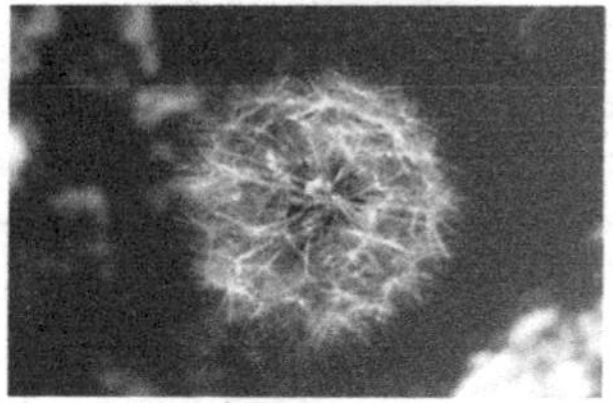

TABLE 4.2 *(Continued)*

Herbal drugs	Synonym	Biological source (family)	Parts of plant use	Traditional diuretic and other uses	Images of herbal plant
Horsegram	Macrotyloma legume	*Macrotyloma uniflorum* (Fabaceae)	Seeds	Diuretic, analgesics	
Watermelon	Citrullus Lanata	*Citrullus lanatus* (Cucurbitaceae)	Fruits	Diuretic, antiulcer, antioxidant	
Italian millet	Kanguni, foxtail millet, Setaria Italica	*Setania indica* (Poaceae)	Stalk and grains	Diuretic, astringent	

TABLE 4.2 *(Continued)*

Herbal drugs	Synonym	Biological source (family)	Parts of plant use	Traditional diuretic and other uses	Images of herbal plant
Black tea	Tea	*Camellia sinensis* Linn (Theaceae)	Fresh leaves	Diuretic	
Coriander	Dhaniya, dhana, malli	*Coriandrum sativum* Linn (Umbeliferae)	Whole plant	Diuretic, antidiabetic, anthelmintic	
Fennel	Fennel	*F. vulgare* (Apiaceae)	Whole plant	Diuretic, antifungal, antibacterial, antioxidant	

4.5 CONCLUSION

The top information is utilized to produce the data regarding the properties and the uses of the medicative plants toward the employment of natural plant as diuretic agents have been comprehensively highlighted. Nowadays, diuretics are the first line medical care for the treatment of high blood pressure, failure, etc. Although, allopathic medicines are obtainable for the treatment of the various excretory organ diseases, high blood pressure, and alternative connected diseases; however, they have numerous adverse effects and contraindications. To beat these adverse effects of artificial medication, the world shifted toward natural remedies. The therapeutic result of the natural herbs is also less pronounced than artificial medication; however, the chances of adverse effects are minimum. These natural diuretics have gained importance due to their low price, greater safety limit, and effectiveness.

KEYWORDS

- **natural**
- **traditional**
- **diuretics**
- **antihypertension**
- **pharmacology**
- **kidney**
- **urine**

REFERENCES

1. Roush, G. C.; Kaur, R.; Ernst, M. E. Diuretics: A Review and Update. *J. Cardiovasc. Pharmacol. Ther.* **2014,** *19*(1), 5–13.
2. Brunton, L. Laurence; Laza, S. John; Parker, L. Keith. Goodman & Gilman's. The Pharmacological Basis of Therapeutics, 11th ed.; The McGraw-Hill Medical Publishing Division: New York, 1941; p 727.
3. Melillo, L. Diuretic Plants in the Paintings of Pompeii. *Am. J. Nephrol.* **1994,** *4*, 423–425.
4. Hober, R. Effect of Some Sulfonamides on Renal Secretion. *Proc. Soc. Exp. Biol. Med.* **1942,** *49*, 87–90.

5. Dutta, K. N.; Chetia, P.; Lahkar, S.; Das, S. Herbal Plants used as Diuretics: A Comprehensive Review. *J. Pharm. Chem. Biol. Sci.* **2014,** *2*(1), 27–32.
6. Wile, D. Diuretics: A Review. *Ann. Clin. Biochem.* **2012,** *49*(5), 419–431.
7. Edemir, B.; Pavenstadt, H.; Schlatter, E.; Weide, T. Mechanisms of Cell Polarity and Aquaporin Sorting in the Nephron. *Pflugers Arch.* **2011,** *46*, 607–621.
8. Maddox, D. A.; Gennari, F. J. The Early Proximal Tubule: A High-Capacity Delivery-Responsive Reabsorptive Site. *Am. J. Physiol.* **1987,** *252*, F573–F584.
9. Chou, C. L.; Nielsen, S.; Knepper, M. A. Structural-Functional Correlation in Chinchilla Long Loop of Henle Thin Limbs: A Novel Papillary Subsegment. *Am. J. Physiol.* **1993,** *265*, F863–F874.
10. Ganong, W. F. Review of Medical Physiology, 21st ed.; McGraw-Hill Professional: New York, 2003; p 720.
11. Barrett, K. E.; Barman, S. M.; Boitano, S.; Brooks, H. *Ganong's Review of Medical Physiology*, 23rd ed.; McGraw Hill Medical: New York, 2009.
12. Kellick, K. A. Diuretics. *AACN Clin. Issues Crit. Care Nurs.* **1992,** *3*, 472–482.
13. Roush, G. C.; Sica, D. A. Diuretics for Hypertension: A Review and Update. *Am. J. Hypertens.* **2016,** *29*(10), 1130–1137.
14. Tripathi, K. D. *Essentials of Medical Pharmacology*, 7th ed.; Jaypee Brothers Medical Publishers Ltd: New Delhi, 2013; pp 579–592.
15. Sharma, H. L.; Sharma, K. K. *Principles of Pharmacology*, 2nd ed.; Paras Medical Publisher: Putlibowli, Hyderabad, 2013; pp 223–237.
16. Kumar, B. S.; Swamy, B. V.; Archana, S.; Anitha, M. A Review on Natural Diuretics. *Res. J. Pharm. Biol. Chem. Sci.* **2010,** *1*(4), 615–634.
17. Satoskar, S. R.; Bhandarkar, D. S.; Rege, N. Nirmala. Pharmacology & Pharmacotherapeutics, 19th ed.; Popular Prakashan private Ltd.: Mumbai, 2005; pp 547–560.
18. Snigdha, M.; Kumar, S. S.; Sharmistha, M.; Lalit, S.; Tanuja, S. An Overview on Herbal Medicines as Diuretics with Scientific Evidence. *Sch. J. Appl. Med. Sci.* **2013,** *1*, 209–214.
19. Sharma, A.; Sharma, A. R.; Singh, H. Phytochemical and Pharmacological Profile of Abutilon Indicum L. Sweet: A Review. *Int. J. Pharm. Sci. Rev. Res.* **2013,** *20*(1), 120–127.
20. Dixit Praveen, K.; Mittal, S. A Comprehensive Review on Herbal Remedies of Diuretic Potential. *Int. J. Res. Phar. Sci. 2013, 3*(1), 41–51.
21. Srivastav, S.; Singh, P.; Jha, K. K.; Mishra, G.; Srivastav, S.; Karchuli, S. M.; Khosa, R. R. Diuretic Activity of Whole Plant Extract of Achyranthes aspera Linn. *Eur. J. Exp. Bio.* **2011,** *1*(2), 97–102.
22. Sundaresan, N.; Ramalingam, R. Pharmacognosy of Azima Tetracantha Lam: A Review. *Int. J. Ayur. Pharm. Res.* **2015,** *3*(12), 13–19.
23. Sahu, A. N.; Damiki, L.; Nilanjan, G.; Dubey, S. Pharmacological Review of Boerhaavia Diffusa Linn. (Punarnava). *Offici. Publi. Phcog. Rev.* **2008,** *2*(4), 14–22.
24. Hook, Mcgee A.; Henaman, M. Evaluation of Dandelion for Diuretic Activity and Variation in Potassium Content. *Int. J. Pharmacog.* **1993,** *31*(1), 29–34.
25. Ashraf, J.; Ghousia, S.; Ahmed, S.; Hasan, M. M. Analgesic, Anti-Inflammatory and Diuretic Activities of Macrotyloma Uniflorum (Lam) Verdc. *Pak. J. Pharm. Sci.* **2018,** *31*(5), 1859–1863.

26. Khan, Z.; Ahmad, N.; Hasan, N.; Srivastava, V.; Ahmad, A.; Yadav, A. Phytopharmacological Study of Citrullus Lanata: A Review. *World J. Pharm. Res.* **2016,** *5*(12), 1289–1300.
27. Nadeem, F.; Ahmad, Z.; Wang, R.; Han, J.; Shen, Q.; Chang, F.; Diao, X.; Zhang, F.; Li, X. Foxitail Millet [Setariaitalica (L.) Beauv.] Grown under Low Nitrogen Shows a Smaller Root System, Enhanced Biomass Accumulation, and Nitrate Transporter Expression. *Front. Plant. Sci.* **2018,** *9*(205), 1–12.
28. Abeywickrama, W. R. K.; Ratnasooriya, D. W.; Amarakoon, T. M. A. Oral Diuretic Activity of Hot Water Infusion of Sri Lankan Black Tea (Camellia sinesis L.) in Rats. *Pharmacogn. Mag.* **2010,** *6*(24), 271–277.
29. Momin, H. A.; Acharya, S. S.; Gajjar, V. A. Coriandrum Sativum-Review of Advances in Phytopharmacology. *Int. J. Pharm. Sci. Res.* **2012,** *3*(5), 1233–1239.
30. Arzoo, Parle M. Fennel: A Brief Review. *Eur. J. Pharm. Med. Res.* **2017,** *4*(2), 668–675.
31. Rajagopal, R. R.; Koumara, V. K. Abutilon indicum L (Malvaceae)-Medicinal Potential Review. *Pharmacog. J.* **2015,** *7*(6), 330–332.
32. Mondal, S.; Parhi, R. P. S.; Dash, G. K. Studies on Diuretic and Laxative Activity of Acacia Suma (Roxb) Barks. *Int. J. Res. Ayurv. Pharm.* **2010,** *1*(2), 510–514.
33. Jahan, N.; Ahmad, R.; Hussain, F. Evaluation of Diuretic Activity of Achyranthes aspera (Chirchita) in Goats. *Pak. Vet. J.* **2002,** *22*(3), 1–4.
34. Shankar, D. G. M.; Gowrishankar, L. N. Priliminary Phytochemical and Diuretic Potential of Methanolic Extract of Azima Tetra Cantha Lam., Leaf. *Int. J. Pharm. Ind. Res.* **2011,** *1*(4), 275–278.
35. Mishra, S.; Aeri, V.; Gaur, K. P.; Jachak, M. S. Phytochemical, Therapeutic, and Ethnopharmacological Overview for a Traditionally Important Herb: Boerhavia diffusa Linn. *Bio. Med. Res. Int.* **2014,** 1–19.
36. Bevin, A. C.; Richard, S. C.; Kevin, S. The Diuretic Effect in Human Subjects of an Extract of Taraxacum Officinale Folium over a Single Day. *J. Alter. Complement Med.* **2009,** *15*(8), 929–934.
37. Kaundal, P. S.; Sharma, A.; Kumar, R.; Kumar, V.; Kumar, R. Exploration of Medicinal Importance of an Underutilized Legume Crop, Macrotyloma Uniflorum (Lam.) Verdc. (Horse Gram): A Review. *Int. J. Pharm. Sci. Res.* **2019,** *10*(7), 3178–3186.
38. Gul, S.; Rashid, Z.; Sarwer, G. Citrullus Lanatus (Watermelon) as Diuretic Agent: An in vivo Investigation on Mice. *Am. J. Drug. Del. Therap.* **2014,** *1*(4), 89–92.
39. Agrawal, S. S.; Paridhavi, M. Herbal Drug Technology; University Press (India) Private Ltd.: Himayatnagar, Hyderabad, 2007; p 317.
40. Wright, C. I.; Van-Buren, L.; Kroner, C. I.; Koning, M. M. Herbal Medicines as Diuretics: a Review of the Scientific Evidence. *J. Ethnopharmacol.* **2007,** *114*(1), 1–31.
41. Mahendra, P.; Bisht, S. Coriandrum Sativum: A Daily Use Spice with Great Medicinal Effect. *Pharmcog. Jour.* **2011,** *3*(21), 84–88.
42. Rather, A. M.; Dar, A. B.; Sofi, N. S.; Bhat, A. B.; Qurishi, A. M. Foeniculum vulgare: A Comprehensive Review of its Traditional Use, Phytochemistry, Pharmacology, and Safety. *Arab. J. Chem.* **2016,** *9*, 1574–1583.

CHAPTER 5

Reviewing the Available Herbal Resources for Treating Psoriasis: Safe and Alternative Way for Therapeutics

SHRUTI DONGARE[1], VAIBHAV SHENDE[1], and
DEBARSHI KAR MAHAPATRA[2*]

[1]*Department of Pharmaceutics, Gurunanak College of Pharmacy and Technical Institute, Nagpur 440026, India*

[2]*Department of Pharmaceutical Chemistry, Dadasaheb Balpande College of Pharmacy, Nagpur 440037, India*

[*]*Corresponding author. E-mail: mahapatradebarshi@gmail.com*

ABSTRACT

Psoriasis is an autoimmune-mediated disease, one of the chronic skin disorders, which has no permanent cure. It is characterized by itchiness, skin rashes, and red scalp. Different types of psoriasis are reported in medical literature. A large variety of artificial medicine agents have additionally been identified to cause skin disorder but their adverse results often challenge the benefit ratio. Therefore, the need for safer medication with simple suitableness is seldom needed. This chapter aims to explore the possible utilization, safety, and therapeutic perspectives of medicinal plant for treating psoriasis and skin diseases which will be a potent, safe, and reliable medical aid in daily life.

5.1 INTRODUCTION

Psoriasis is an autoimmune inflammatory disorder that has results in an enhanced proliferation of the vegetative cell. The skin becomes unquiet and painful to nearly 1.5–2% of the world population. Normally, the cells of the skin die and are replaced by new cells. However, in the case of psoriatic condition, the cells grow at a quicker rate and appear like patches rising on the skin. It is one of the papules squamous skin disorders.[1,2] The most common symptoms of this skin problem include irritation, dryness, redness, flaky patches of skin, genital sores, pain in joint, thickening and browning of nails, and severe dandruff.[3,4,5] It is often classified clinically as plaque psoriasis, gultate psoriasis, fexual psoriasis, pustular psoriasis, erythrodermic psoriasis, hail psoriasis, and psoriatic arthritis.

A clinical diagnosis of any psoriasis case is based upon the nature of the skin and its characterization, which is generally done by a clinical examination. At present day, there is no precise test available to diagnose this skin problem; however, general diagnostic test could also be performed to differentiate it from mycosis. The diagnosis of the affected joints may be done by X-ray scanning methodology. Modern-day psoriasis diagnosis may include dermatological conditions such as discoid eczema, seborrheic, and so on.[6,7,8]

5.2 MANAGEMENT OF PSORIASIS

The symptoms of psoriasis can be managed nonpharmacologically by regular diet and water intake only. However, topical (coal tar, anthralin, calcipotriene, tazarotene, and tacrolimus) (Table 5.1) and synthetic (methotrexate, acitretin, and cyclosporin) (Table 5.2) therapies for the pharmacological treatment of psoriasis have been reported.

TABLE 5.1 Topical Therapies of Psoriasis Treatment.

Drug	Brand name	Dose
Coal tar	Psoriasin	Use one or two times a day
Anthralin	Dianthrol	0.1% concentration is applied once a day and washed off thoroughly after 10 min–1 h
Calcipotriene	Daivonex	Cream, 0.005%. Apply a thin layer of cream to the affected skin twice daily and rub it gently and completely
Tazarotene	Tazorate forte	Topical cream 0.1%
Tacrolimus	Acroil forte	0.03% ointment; two times a day

TABLE 5.2 Systemic Therapies of Psoriasis Treatment.

Drug	Brand name	Dose
Methotrexate	Altrex	Single dose: 7.5 mg/week orally, IM, IV
Acitretin	Acret	25 mg every day to maintenance dose of 50 mg daily
Cyclosporin	Graftin	3–5 mg/kg given orally in two divided dose

5.3 HERBS USED IN THE TREATMENT OF PSORIASIS

Mother Nature has several hidden treasures for safe and natural way of treating psoriasis. Plants such as *Capsicum annuum* **(1)**, *Wrightia tinctoria* **(2)**, *Curcuma longa* **(3)**, *Cassia tora* **(4)**, *Allium cepa* **(5)**, *Aloe barbadensis* **(6)**, *Jasminum polyanthus* **(7)**, *Matricaria recutita* **(8)**, *Mahonia aquifolium* **(9)**, *Momordi cacharantia* **(10)**, *Azadirachta indica* **(11)**, *Thespesia populnea* **(12)**, *Tribulus terristris* **(13)**, *Silibum marianum* **(14)**, *Calendula officinalis* **(15)**, *Berberis vulgaris* **(16)**, *Allium sativum* **(17)**, *Nigella sativa* **(18)**, and *Smilax china* (**Figure 5.1**) have been identified as potential ethnobotanical reserves for treating psoriasis and related skin problems.

FIGURE 5.1 Some common promising herbal components for treating psoriasis.

5.3.1 *CAPSICUM ANNUUM (FAMILY: SOLANACEAE; COMMON NAME: RED PEPPER, CHILI PEPPER)*

It is normally referred to as cayenne, its chief element being chemical irritant. One hypothesis on the morbific method of malady of the skin suggests a tissue inflammatory etiology mediate through substance-P (SP). SP activates inflammatory cells and ultimately perpetuates that psoriatic lesions are acquainted to be extra-densely inverted with higher SP content than management or uninvolved psoriatic skin. The chemical irritant stimulates the discharge of SP by binding to the vanilloid receptor on slow conducting, unmyelinated nerve cell, and ultimately results in its depletion. The substances present in this plant have been identified as key ingredients in treating psoriasis.[9,10,11]

5.3.2 *WRIGHTIA TINCTORIA (FAMILY: APOCYNACEAE; COMMON NAME: SWEET INDRAJAO, PALA INDIGO PLANT)*

The hydroalcholic extract of *W. tinctoria* leaves has shown vital antipsoriatic result on mouse-tail model as compared to isoretionic acid. It has been found that the extract, which made vital activity against orthokeratosis, acts as a distinguished inhibitor activity in DPPH assay and nitric oxide.[12]

5.3.3 *CURCUMA LONGA (FAMILY: ZINGIBERACEAE; COMMON NAME: TURMERIC)*

C. longa, normally called turmeric, is a stem herb that grows to a height of 3–5 ft. It has oblong, pointed leaves, and bears funnel-shaped yellow flowers piping out of large brackets. The rhizome is the portion of the plant used medicinally. It is additionally reported to have phK activity by the curcumin and calciportriol which corresponded to the severity of parakeratosis and effectively decreases the keratinocyte transferrin receptor expression and density of epidermic CD8+ T cells in case of skin diseases and psoriasis.[10]

5.3.4 *CASSIA TORA (FAMILY: LEGUMINOSAE; COMMON NAME: CHAKUNDA)*

It has pinnate leaves that are 10-cm long and every leaf has three pairs of leaflets that are opposite, ovate, rectangular, and oblique at the bottom. This plant, also called charota, chakunda, and reaping hook shrub domestically, has been historically used for the treatment of psoriasis, skin disorders, and different skin ailments. *C. tora* leaves, rich in organic compounds, additionally contained aloe emodin, which can be useful for the treatment of skin diseases.[13]

5.3.5 *ALLIUM CEPA (FAMILY: LILIACEAE; COMMON NAME: ONION)*

A. cepa is most well-liked by patients with seborrheic disease of the skin and psoriasis. Following any excision, the onion extract gel enhances the healing rate of the scar by improving and promoting smoothness, reducing redness, and reducing inflammatory manifestation at the excision site after 4–6-week treatment. The cumulative effort has applications in treating psoriatic symptoms.[14]

5.3.6 *ALOE BARBADENSIS (FAMILY: LILIACEAE; COMMON NAME: ALOE VERA)*

Aloe vera is a plant proverbial for its soothing and cooling properties. This plant (in the form of cream containing a minimum of 0.5% aloe, applied up to three times per day for 5 days) also has applications in treating psoriasis, skin ailments and also facilitates regeneration of skin cells. It also heals skin irritation by decreasing the inflammation and facilitates to clear a light skin disorder flare.[15]

5.3.7 *JASMINUM POLYANTHUS (FAMILY: OLEACEAE; COMMON NAME: CLIMBING JASMINE, WHITE JASMINE)*

Jasmine flowers are employed in the treatment of a variety of skin diseases. The paste created by crushing the flowers is directly applied on the affected components of the skin and this is found to decrease the pain and, therefore, the skin sensation of psoriatic inflammations.

5.3.8 MATRICARIA RECUTITA (FAMILY: ASTERACEAE; COMMON NAME: CHAMOMILLA, KAMILLA)

Traditionally, *M. recutita* is used in several therapies on skin rejuvenation, stress relief, and so on and is also employed in GI disorders. The byproduct of matricin is a non-oil extract component that is taken into medical applications for psoriasis and skin diseases. It is known to possess selective leukotriene B4 (LTB4) inhibitory potential.[16,17]

5.3.9 MAHONIA AQUIFOLIUM (FAMILY: BERBERIDACEAE; COMMON NAME: OREGON GRAPE)

M. aquifolium is an awfully standard plant employed in skin disorders, particularly in psoriatic plaques. The impact of bark extract of Oregon grape, and its main constituents, has lipoxygenase and lipid-peroxidation activity reducing functions. It is also associated in nursing substance of keratinocytes growth. The benzlisoquinoline organic compounds berbamine and oxyacanthine are imperative inhibitors for treating psoriasis and skin diseases.[18,19,20,21]

5.3.10 MOMORDI CACHARANTIA (FAMILY: CUCURBITACEOUS; COMMON NAME: BITTER GOURD)

M. cacharantia is regionally referred to as bitter gourd. It is associate nonwoody, tendril-bearing tracheophyte, growing up to 5-m high, straight, and 3–12 cm across. It is utilized as boiling, infusions, decoctions, tinctures, and infusions to treat microorganism infections and conjointly claimed to be a good against sort of skin condition like psoriasis, acne, and wounds. It is a common Indian herb having varied healthy properties for the treatment of various sickness, has wide usage in wound healing, plays a role in opposing-fungal, and remarkable antidiabetic activity. This herb has been reported for its quality treatment of leprosy and psoriasis.[22,23]

5.3.11 AZADIRACHTA INDICA (FAMILY: MELIACEAE; COMMON NAME: NEEM)

The alternate leaves are 20–40 cm long, with 20–31 medium to dark inexperienced leaflets about 3–8 cm long. The stem bark is burnt and ash is applied locally on boils. Stewing of leaves is employed to bathtub for the treatment of body infection, including psoriasis. Its stewing is also taken orally for the treatment of skin disorders. The seed oil is employed outwardly to kill hair lice and to treat dandruff.[24]

5.3.12 THESPESIA POPULNEA (FAMILY: MALVACEAE; COMMON NAME: PORTIA TREE)

Botanical name of Indian tulip tree puvarasuits is *Thespesia populnea* which belongs to family Malvaceae. Historically, it is claimed to be helpful for the treatment of cutaneal affections like itch, mycosis disease of the skin, disease of the skin, and herpetic disease. The therapeutic oil, prepared by boiling the bottom bark in the oil, is applied outwardly in the psoriasis, various diseases of the skin, and itching. *T. populnea* bark extract on Perry's scientific mouse-tail model has been found to enhance orthokeratosis. The plant has been reported to contain carbohydrates, glycoside, tannins, flavonoids, terpenoids, proteins, and fixed oil which contribute to main therapeutic activities.[25]

5.3.13 TRIBULUS TERRISTRIS (FAMILY: ZYGOPHYLLACEAE; COMMON NAME: GOKHARU)

T. terristris, normally called puncture vascular plant, has an extended history of use throughout the globe. It has been employed in China for last four decades to treat various disease conditions like ejaculation, psoriasis, liver disease, and skin problem. Other ancient Eastern cultures have ethnobotanical applications in treating psoriasis and numerous skin ailments.[26]

5.3.14 SILIBUM MARIANUM (FAMILY: ASTERACEAE; COMMON NAME: MILK THISTLE)

S. marianum, normally referred to as milkweed or vishnukranti. The leaves are either lobate or pinnate with prickly edges. They are rectangular to lance like, hairless, shiny inexperienced, with achromatic veins. It triggers a pandemic of disease of the skin, including psoriasis and inbound cases of guttate disease of the skin. The milkweed has been shown to inhibit human T-cell activation which takes place in disease of the skin.[27]

5.3.15 CALENDULA OFFICINALIS (FAMILY: ASTERACEAE; COMMON NAME: MARIGOLD)

C. officinalis are a number of the quite common Indian herbs having numerous medicative properties for the treatment of various kinds of diseases such as antifungal, wound healing, and antidiabetic agent. *C. officinalis* are short-lived aromatic nonwoody perennial herbs; the leaves are rectangular lance-like, 5–17 cm long, furry on each aspect, and with margins entire or sometimes waved or decrepit toothed. This herb is ethically known for its utility in various simmering, infusions, and tinctures. In traditional system of medicines, it has been employed for treating psoriasis and skin diseases of infectious origin.[22,23]

5.3.16 BERBERIS VULGARIS (FAMILY: BERBERIDACEAE; COMMON NAME: BARBERRY)

Barberry is available in capsules, tea, or tinctures. It is used as associate degree inhibitor, antipsoriatic, anti-inflammatory, and apparently prevents poison formation within the internal organ. Berberine is a lively compound within this herb, that is employed for treating skin condition and psoriatic symptoms, which is less attackable than chloromycetin. It is conjointly used as antibiotic to kill or stop the expansion of the microorganism. It causes symptoms like dysentery, tract infection, and epidemic cholera.[28]

5.3.17 ALLIUM SATIVUM (FAMILY: AMARYLLIDACEOUS; COMMON NAME: GARLIC)

Garlic oil can be applied to the skin to treat a multitude of skin ailments, including psoriasis due to its high anti-inflammatory properties. It can also relief itchy psoriasis and outbreaks on the skin.[29]

5.3.18 NIGELLA SATIVA (FAMILY: RANUNCULACEARE; COMMON NAME: BLACK CUMIN, BLACK CARAWAY)

Tazarotene gel was applied as a standard in the study to screen the antipsoriatic activity of *N. sativa* Linn seeds, which produced a massive way for the treatment. It was studied by performing mouse-tail model and by in-vitro methods using cell lines which demonstrated multiple applications.[30,31,32]

5.3.19 SMILAX CHINA (FAMILY: SMILACACEAE; COMMON NAME: CATBERIERS, GREENBERIES)

The presence of quercetin in the rhizomes of *S. china* was found to possess antipsoriatic activity. There was a decrease in the thickness of the epidermis with a marked decrease in the migration of the leukocyte.[33,34]

5.4 CONCLUSION

This chapter will positively open new pharmacotherapeutic perspectives for the better management as well as treatment of psoriasis and symptoms of associated skin disorders in modern context by using the natural plant components such as *C. annuum*, *W. tinctoria*, *C. longa*, *C. tora*, *A. cepa*, *A. barbadensis*, *J. polyanthus*, *M. recutita*, *M. aquifolium*, *M. cacharantia*, *A. indica*, *T. populnea*, *T. terristris*, *S. marianum*, *C. officinalis*, *B. vulgaris*, *A. sativum*, *N. sativa*, and *S. china* where synthetic molecules have shown poor pharmacodynamics and pharmacokinetic properties.

KEYWORDS

- **psoriasis**
- **skin**
- **symptoms**
- **herbal**
- **treatment**
- **therapy**
- **natural**

REFERENCES

1. Chandrasekar, R.; Sivagami, B. Alternative Treatment for Psoriasis—A Review. *Int. J. Res. Dev. Pharmacy Life Sci.* **2016,** *4*, 2188–2197.
2. de Korte, J.; Mirjam Sprangers, A. G.; Femke Mombers, M. C.; Jan Bos, D. Quality of Life in Patients with Psoriasis: A Systemic Literature Review. *J. Invest. Dermatol. Symp. Proc.* **2014,** *9*, 140–147.
3. Fujii, R.; Mould, J.; Tang, B. PSY46 Burden of Disease in Patient with Diagnosed Psoriasis in Brazil: Result from 2011 NHWS. *Value Health* **2012,** *15*, A107.
4. Russo, P. A.; Ilchef, R.; Cooper, A. J. Psychiatric Morbidity in Psoriasis: A Review. *Aust. J. Dermatol.* **2004,** *45*, 155–161.
5. Samponga, F.; Taboli, S. Living with Psoriasis: Prevalence of Shame, Anger, Worry, and Problem in Daily Activities and Social Life. *Acta Dermatol. Venereol.* **2012,** *92*, 299–303.
6. Weigle, N. Psorasios. *Am. Fam. Physican* **2013,** *87*, 626.
7. Johnson, M. A. N.; Armstrong, A. W. Clinical and Histologic Diagnostic Guidelines for Psoriasis: A Critical Review. *Clin. Rev. Allergy Immunol.* **2013,** *386*, 1137.
8. Mease, P. J.; Armstrong, A. W. Managing Patients with Psoriatic Disease: The Diagnosis and Pharmacologic Treatment of Psoriatic Arthritis in Patients with Psoriasis. *Drugs* **2014,** *74*, 423–441.
9. Bernstein, J. E.; Parish, L. C. Effects of Topically Applied Capsaicin on Moderate and Severe Psoriasis Vulgaris. *J. Am. Acad. Dermatol.* **1986,** *15*, 504–507.
10. Joe, B.; Lokesh, B. R. Effect of Curcumin and Capsaicin on Arachidonic Acid Metabolism and Lysomal Enzyme Secretion by Rat Peritoneal Macrophages. *Lipids* **1997,** *32*, 1173–1180.
11. Llis, C. N.; Berberian, B.; Sulica, V. I.; Dodd, W. A. A Double-Blind Evaluation of Topical Capsaicin in Pruritic Psoriasis. *J. Am. Acad. Dermatol.* **1993,** *29*, 438–442.
12. Dhanabal, S. P.; Raj, B. A.; Muruganantham, N. Screeing of *Wrightia tinvtoria* Leaves for Anti-psoriatic Activity. *Hyg. J. Drugs Med.* **2012,** *4*, 73–78.
13. Singhal, M.; Kansara, N. *Cassia tora* L. Creams Inhibit Psoriasis in Mouse Tail Model. *Pharm. Crops* **2012,** *3*, 1–6.

14. Shams-Ghahfarikhi, M.; Shokoohamiri, M.-R. In-vitro Antifungal Activities of *A. cepa, Allium sativum* and Ketoconazole against Some Pathogenic Yeast and Dermatophytes. *Fitoterapia* **2006,** *77*, 321–325.
15. Choonhakarn, C.; Busaracome, P.; Sripanidkulchai, B. A Prospective, Randomized Clinical Trail Comparing Topical Aloe Vera with .01% Triamcinolone Acetonide in Mild to Moderate Plaque Psoriasis. *J. Eur. Acad. Dermatol. Venereol.* **2010,** *24*, 168–172.
16. Murti, K.; Panchal, M. A.; Gajera, V. Pharmacological Properties of *Matricaria recutia*: A review. *Pharmacologia* **2012,** *3*, 348–351.
17. Safayhi, H.; Sabieraj, J.; Sailer, E. R.; Ammon, H. P. Chamazulene: An Antioxidant—Type Inhibitor of Leukotriene B_4 Formation. *Planta Med.* **1994,** *60*, 410–413.
18. Galle, K.; Müller-Jakic, B. Analytical and Pharmacological Studies on *Mahonia aquifolium*. *Phytomedicine* **1994,** *1*, 59–62.
19. Gulliver, W. P.; Donsky, H. J. A Report on Three Recent Clinical Trials Using *Mahonia aquifolium* 10% Topical Cream and the Review of the Worldwide Clinical Experience with *Mahonia aquofolium* for the Treatment of Plaque Psoriasis. *Am. J. Ther.* **2005,** *12*, 398–406.
20. Misik, V.; Bezakova, L.; Malekova, L. Lipoxygenase Inhibition and Antioxidant Properties of Protoberberine and Atropine Alkaloids Isolated from *Mahonia aquifolium*. *Planta Med.* **1995,** *61*, 372–373.
21. Muller, K.; Ziereis, K. The Antipsoriatic *Mahonia aquifolium* and Its Active Constituents, I. Pro and Antioxidant Properties and Inhibition of 5-Lipoxygenase. *Planta Med.* **1994,** *60*, 421–424.
22. Brown, D. J.; Dattner, A. M. Medical Journal Article on Herbs for Common Skin Condition *Arch. Dermatol.* **1998,** *134*, 1401–1404.
23. Roopashree, T. S.; Dang, R.; Shobha Rani, R. H. Antibacterial Activity of Anti-psoriatic Herbs: *Cassia tora, Momordica cacharantia* and *Calendula officinalis*. *Int. J. Appl. Res. Nat. Prod.* **2008,** *1* (3), 20–28.
24. Mundada, A. S.; Mahajan, M. S.; Gangurde, H. H.; Borkar, V. S.; Gulecha, V. S.; et al. Formulation and Evaluation of Polyherbal Anti-psoriatic. *Pharmacol. Online* **2009,** *2*, 1185–1191.
25. Vijayalakhsmi, A.; Ravichandiran, V.; Malarkadi, V.; Nirmala, S.; Jaykumari, S. Screening of Flavonoid Quercetin from the Rhizome of *Smilax china* Linn. for Anti-psoriatic Activity. *Asian Pac. J. Trop. Biomed.* **2012,** *2*(4), 269–275.
26. Rajesh, B. N.; Fleming, A.; Devada, S.; Ranvir, R.; Sundar, R. Anti-psoriatic Effect of *Tribulus terrestris* Extract by Topical Application in Mouse Model of Contact Dermatitis. *Int. J. Vet. Sci.* **2013,** *2*(1), 7–11.
27. Sabir, S.; Arsshad, M.; Asif, S.; Chaudhari, S. K. An Insight into Medicinal and Therapeutic Potential of *Silybum marianum* (L.) Gaertn. *Int. J. Biosci.* **2014,** *4* (11), 104–115.
28. Rao, S. G.; Udupa, A. L.; Udupa, S. L.; Rao, P. G. M.; Rao, G.; Kulkarni, D. R. Calendula and Hypericum: Two Homeopathic Drugs Promoting Wound Healing in Rats. *Fitoterapia* **1991,** *62*, 508–510.
29. Das, I.; Saha, T. Effect of Garlic on Lipid Peroxidation and Antioxidation Enzymes in DMBA-Induced Skin Carcinoma. *Nutrition* **2009,** 25, 459–471.

30. Chun, H.; Shin, D. H.; Hong, B. S.; Cho, W. D.; Cho, H. Y.; et al. Biochemical Properties of Polysaccharides from Black Pepper. *Biol. Pharm. Bull.* **2002,** *25* (9), 1203–1208.
31. Ghosheh, O. A.; Houdi, A. A.; Crooks, P. A. High Performance Liquid Chromatographic Analysis of the Pharmacologically Active Quinones and Related Compounds in the Oil of the Black Seed (*Nigella sativa* L.). *J. Pharm. Biomed. Anal.* **1999,** *19* (5), 757–762.
32. Dwarampudi, L. P.; Palaniswamy, D.; Nithyanantham, M.; Raghu, P. S. Anti-psoriatic Activity and Cytotoxicity of Ethanolic Extract of *Nigella Sativa* Seed. *Pharmacogn. Mag.* **2012,** *8* (32), 268–272.
33. Anonymous (1072) National Institute of Science Communication CSIR. *Wealth of India* **n.d.,** *4*, 366.
34. Vijayalakshmi, A.; Ravichandirn, V.; Malarkodi, V.; Nirmala, S.; Anusha, M.; et al. Anti-psoriatic Activity of *Smilax china* Linn. Rhizome. *Indian J. Pharm. Educ. Res.* **2013,** *47* (1), 82–89.

CHAPTER 6

Nutraceuticals and Brain Disorders

AKSHADA ATUL BAKLIWAL, VIJAY SHARADKUMAR CHUDIWAL, and SWATI GOKUL TALELE*

Department of Pharmaceutics, Sandip Institute of Pharmaceutical Sciences, Nashik, India

**Corresponding author. E-mail: swatitalele77@gmail.com*

ABSTRACT

The term "Nutraceuticals" was derived from "Nourishment" and "Pharmaceuticals" by Stephen Defelice in 1989. For various sorts of illnesses, the nutraceuticals are an elective treatment. Nutraceutical way to deal with this sickness is a promising system, particularly in certain regions, it is more alluring than others. In this, we center around neuroissues like Parkinson's sickness, mental imbalance, neurodevelopmental disorders, and their belongings. In this situation, natural products (nutraceuticals) assume an imperative job which is plant based. This chapter basically analyzes the role of nutrients and cofactors, dietary adjustments and gut anomalies, probiotics and prebiotics, phytochemicals, and ecological factors so as to decide the condition of proof in nutraceutical-based disease management practices.

6.1 HISTORY

The concept of nutraceuticals returned 3000 years ago. In the mid-1900s, the US food manufacturer started adding little amount of iodine to salt to anticipate goiter. In Japan, England, and different nations, nutraceuticals are as of now winding up some portion of dietary scene; these days nutraceuticals are the most quickly developing fragments of the business and the worldwide nutraceutical market is assessed as USD 117 billion.[1–4]

6.2 INTRODUCTION

The term "nutraceuticals" was derived from "Nutrition" and "Pharmaceuticals" by Stephen Defelice, who is the founder and executive of establishment for imaginative prescription. The real utilization of nutraceuticals is to accomplish attractive helpful results with diminished symptoms. Around 2000 years prior, Hippocrates accentuated "let nourishment be your prescription and drug be your nourishment's." Nutraceuticals are utilized as nourishment or part of nourishment which will give medicinal or medical advantages including avoidance or treatment of infection. Home-grown nutraceuticals are amazing instruments in keeping up wellbeing and act against healthfully prompted intense and incessant illnesses by advancing ideal wellbeing, life span, and personal satisfaction.[5]

6.2.1 UTILIZATION OF NUTRACEUTICALS IN THE TREATMENT OF DIFFERENT DISEASES[6–10]

6.2.1.1 RICE BRAN AND CARDIOVASCULAR DISEASES, EYE SIGHT

Rice bran brings down the serum cholesterol levels in the blood, brings down the degree of low-density lipoproteins (LDL) and expands the level of high-density lipoproteins (HDL) in cardiovascular wellbeing. Higher the proportion more will be the danger of coronary heart infections. Rice and wheat contain both lutein and zeaxanthin, which improves visual perception and diminishes the opportunity of waterfalls. The fundamental unsaturated fats, omega-3, omega-6, omega-9, and folic corrosive in rice grain are additionally advancing eye wellbeing.

6.2.1.2 CORN (CARDIOVASCULAR FAILURE, LUNG MALIGNANCY)

Corn's commitment to heart wellbeing lies in its fiber, however, in the critical measures of folate that corn supplies. Corn keeps up the homocysteine, a middle of the road item is a significant metabolic procedure called the methylation cycle. Homocysteine is straightforwardly in charge of harm of vein coronary episode, stroke, or fringe vascular malady. It has been assessed that utilization of 100% of the daily value (DV) of folate would, without anyone else, diminish the quantity of coronary failures endured by 10%. Corn additionally contains cryptoxanthin, a characteristic carotenoid color. It has been discovered that cryptoxanthin can lessen the danger of lung malignant growth by 27% on everyday utilization.

6.2.1.3 DIETARY POLYPHENOLS USED IN THE TREATMENT OF DIABETES

As of late, there is developing proof that plant nourishments polyphenols, because of their organic properties, might be novel nutraceuticals and advantageous medicines for different parts of sort 2 diabetes mellitus. Polyphenolic mixes can likewise anticipate the advancement of long haul diabetes inconveniences including cardiovascular ailment, neuropathy, nephropathy, and retinopathy.

6.2.1.4 SORGHUM (AGAINST PATHOGEN)

Sorghum is the fundamental dietary hotspot for 3-deoxyanthocyanidins, which are available in enormous amounts in the wheat of certain cultivars. The guard instrument of sorghum against pathogen is because of a functioning procedure, bringing about the aggregation of significant levels of 3-deoxyanthocyanidin phytoalexins in tainted tissues.

6.2.1.5 BUCK WHEAT (OBESITY CONSTIPATION)

Buckwheat seed proteins are helpful for weight and blockage acting like regular strands present in nourishment. 5-Hydroxytryptophan and green tea concentrates may advance weight reduction.

6.2.1.6 *BETA-CAROTENE (MALIGNANCY)*

Beta-carotene is the primary wellspring of nutrient and has antioxidant properties which help in avoiding malignant growth and different sicknesses. Among different carotenes, beta-carotene is the most dynamic cell reinforcement. Alpha and beta-carotenes, alongside gamma-carotene and the carotenes, lycopene, and lutein which don't change over to nutrient An, appear to offer security against lung, colorectal, bosom, uterine, and prostate diseases. Beta-carotene is the more typical structure and can be found in yellow, orange, and green verdant foods grown from the ground. These can be carrots, spinach, lettuce, tomatoes, sweet potatoes, broccoli, melon, oranges, and winter squash.

6.2.1.7 *THE TREATMENT OF JOINT PAIN*

Joint pain is a typical ailment where the end-point brings about joint substitution medical procedure. The utilization of nutraceuticals is an elective treatment for obsessive appearances of joint sickness. The viability of fish oils (e.g., cod liver oil) in the eating routine has been exhibited in a few clinical preliminaries, creature-nourishing investigations, and in-vitro models that copy ligament demolition in joint sickness. Other than this, there is some proof of different nutraceuticals, for example, green tea, home-grown concentrates, chondroitin sulfate, and glucosamine.

6.2.1.8 *NUTRACEUTICALS UTILIZED AGAINST ALZHEIMER'S DISEASE*

Alzheimer's disease (AD), likewise called decrepit dementia of the Alzheimer type, essential degenerative dementia of the Alzheimer's sort, or just Alzheimer's, is the most widely recognized type of dementia. The different nutraceuticals, which are utilized to fix Alzheimer's illness are as follows:

(a) Antioxidants: Cell reinforcements like nutrient E and nutrient C.
(b) *Gingko biloba*: *Ginkgo biloba* is maybe the most examined herbs with reference to memory, discernment, generally cerebrum execution, and surely AD.

(c) Huperzine alpha: Huperzine alpha or huperzine A will be an engaging plant exacerbate that is separated from club.
(d) Greenery, or *Huperzia serrata*: It is a sesquiterpene alkaloid, which is an intense and reversible inhibitor of acetylcholinesterase.

6.2.1.9 THE TREATMENT OF EATING REGIMEN-RELATED INFECTIONS

In Western social orders, the frequency of eating regimen related sicknesses is continuously expanding because of more noteworthy accessibility of hypercaloric nourishment and an inactive way of life. Stoutness, diabetes, atherosclerosis, and neurodegeneration are real diet-related pathologies that offer a typical pathogenic denominator of poor quality irritation. Utilitarian nourishments and nutraceuticals may speak to a novel remedial way to deal with anticipate or weaken diet-related ailment in perspective on their capacity to apply mitigating reactions. Specifically, enactment of intestinal T-administrative cells and homeostatic guideline of the gut microbiota can possibly decrease second-rate irritation in eating regimen-related maladies.

6.2.1.10 VISION IMPROVING SPECIALISTS

Lutein is one of the most significant carotenoids found in numerous leafy foods like mangoes, corn, sweet potatoes, carrots, squash, tomatoes, and so forth. Lutein dipalmitate is found in the plant Hele-niumautumnale. Lutein, otherwise called helenien, is utilized for the treatment of visual issues. Zeaxanthin is utilized in customary Chinese medication principally for the treatment of visual issues. Nourishment wellsprings of zeaxanthin incorporate corn, egg yolk, green vegetables, and natural products, for example, broccoli, green beans, green peas, Brussels sprouts, cabbage, kale, collard greens, spinach, lettuce, kiwi, and honeydew. Lutein and zeaxanthin are likewise found in weeds, green growth, and the petals of many yellow blossoms. In green vegetables, foods grown from the ground yolk, lutein, and zeaxanthin exist in nonesterified structures.

6.3 NUTRACEUTICALS WITH SPECIAL EMPHASIS ON BRAIN DISORDERS

6.3.1 DIFFERENT BRAIN DISORDER

Some of the different brain disorders are:

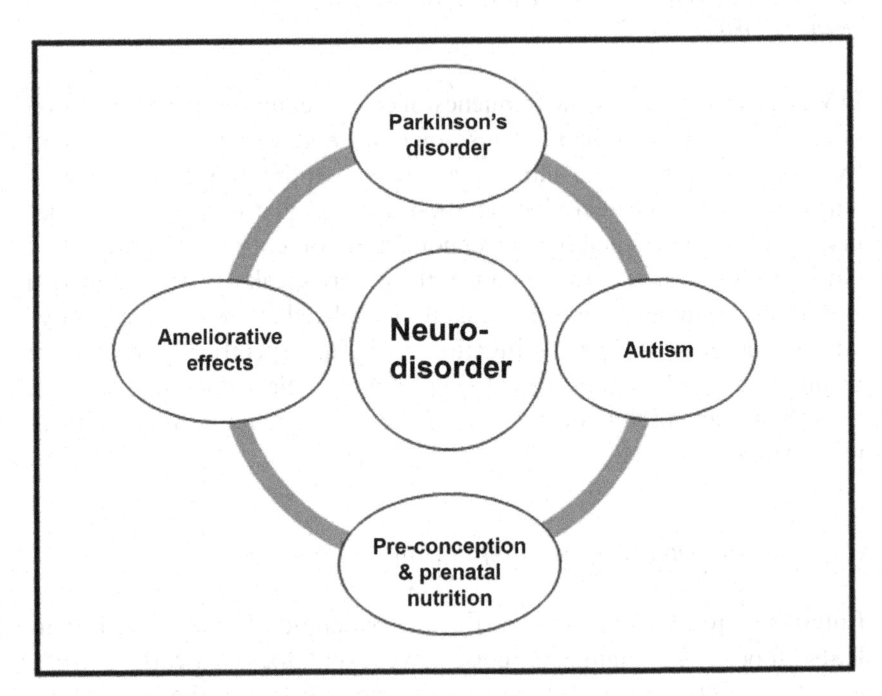

FIGURE 6.1 Different neurodisorders and their effects.

6.3.1.1 PARKINSON'S DISORDER

PD is defined pathologically by the dynamic loss of dopaminergic neurons in the substantia nigra (SN) standards compact adjoined by the nearness of intracellular Lewy bodies. Parkinsonism, alongside parkinsonian disorder, ought to be recognized from PD.[11–15] Parkinsonism is a term that alludes just to the clinical manifestations of PD, for example, the event of tremors and dementia, however, bears no ramifications of infection instrument, while PD alludes to the pathology portrayed previously. The precise components of PD are not yet completely comprehended, albeit a few elements, including protein misfolding, oxidative pressure, and mitochondrial brokenness, have been accounted for. With regards to PD, a few nutraceuticals have been appeared to give neuroprotection in exploratory models and may fill in as options in contrast to manufactured medication mixes like L-DOPA that is known to cause numerous bothersome reactions.[16–19] The systems by which they work can be comprehensively arranged into the accompanying subjects: (1) responsive oxygen species (ROS)/free radical searching, (2) hostile to aggravation, (3) iron chelation, (4) adjustment of cell flagging pathways, (5) against apoptosis, and (6) mitochondrial homoeostasis, albeit a few nutraceuticals basically work by means of a horde of unthinking pathways as opposed to holding fast to a solitary component.[20–23]

6.3.1.2 AUTISM

Autistic spectrum disorders (ASDs) are progressively pervasive neurodevelopmental conduct disorders of weakened verbal and nonverbal correspondence and socialization abilities among youngsters. Mental imbalance and related issues are progressively predominant conduct disorders of debilitated verbal and nonverbal correspondence and socialization attributable to neurodevelopmental variations from the norm. The latest gauge for the predominance of medically introverted issue is about 1% on a worldwide scale. Etiology of mental imbalance is multifactorial and multidimensional that is helpful to entomb vention much harder. Heterogeneity of hereditary variables, oxidative pressure, immune system component, and epigenetic instruments convolute the idea of pathogenesis of the sickness. Nutraceutical way to deal with infection is a promising procedure, particularly in certain regions, it is more alluring than others. This audit fundamentally dissects the jobs of nutrients and cofactors, dietary modifications and gut variations from the norm, probiotics and prebiotics, phytochemicals, and ecological factors so as to decide the condition of proof in nutraceutical-based chemical imbalance the board rehearses.[24–30]

6.3.1.3 AMELIORATIVE EFFECTS OF NUTRACEUTICALS IN NEUROLOGICAL DISORDER

Neurodegeneration is an age-related dynamic weakening of neuronal structures and capacities eventually prompting psychological handicap and dementia. Dementia is anything but a typical piece of aging but instead an aftereffect of neurotic aging of the mind portrayed by the nearness of mental incapacities at different degrees of seriousness. Ayurveda groups mental incapacities at three distinct levels: Dhi (securing/learning), Dhruti (maintenance of memory), and Smruti (review/recovery of memory). Disintegration in any of these procedures might be characterized as dementia. Aging is the premier unavoidable hazard factor for creating age-related neurodegeneration and dementia. Age-subordinate ruinous biochemical processes keep on overstating during aging of the mind, prompting neurodegeneration, and dementia. The most common neurodegenerative issues (NDs) related with dementia incorporate Alzheimer's malady (AD), Parkinson's sickness (PD), awful mind damage, incessant horrendous encephalopathy, stroke, and epilepsy. These neurodegenerative conditions share numerous auxiliary and unthinking shared characteristics at the atomic, cell, and practical levels. One such shared trait is the anomalous protein misfolding seen in NDs. Considering the age-related transaction of heterogeneous cell and subatomic occasions, which when irritated switch ordinary maturing of the cerebrum to the obsessive course prompting NDs and dementia, devouring pleiotropic phytonutrients (e.g., those found in organic products, nuts, vegetables, and flavors) applying cancer prevention agent, mitigating, cholinergic, and intellectual impacts, from at an opportune time during the prodromal age and proceeding all through life, may advance "sound maturing of mind" and hinder the beginning of neurodegenerative subjective decrease and dementia.[31–36]

6.3.1.4 NEURODEVELOPMENTAL DISORDERS, PRECONCEPTION NUTRITION, AND PRENATAL NUTRITION

Preconception and prenatal nutrition are basic for fetal mental health. Be that as it may, its relationship with posterity neurodevelopmental issue is not surely known. This investigation intends to methodically audit the relationship of assumption and prebirth sustenance with posterity danger of neurodevelopmental issue. Neurodevelopmental results included mental imbalance range issue (ASD), a lack of ability to concentrate consistently issue hyperactivity (ADHD) and scholarly inabilities. A sum of 2169 articles was screened, and 20 articles on ASD and 17 on ADHD were in the long run inspected. We found a general backward relationship between maternal folic corrosive or multivitamin supplementation and youngsters' danger of ASD; a meta-examination including six forthcoming accomplice studies assessed a recurrence risk (RR) of ASD of 0.64 (95% CI: 0.46, 0.90). Information on relationship of other dietary elements and ASD, ADHD, and related results were uncertain and warrant future examination. Future investigations should incorporate thorough and increasingly target strategies to measure the healthful exposures and investigate elective examination plan, for example, Mendelian randomization to assess potential causal impacts.[37–40]

Maternal nutrition is basic for fetal mental health. Maternal eating routine preceding pregnancy is significant for improving nourishing status which assumes an imperative job in keeping up a solid pregnancy and supporting the creating baby. Nutrition around the hour of origination is significant for gamete capacity and placental advancement. Beginning 2–3 weeks after preparation, the incipient organism experiences organized procedures of neuronal expansion and relocation, neurotransmitter arrangement, myelination, and apoptosis to build up the fetal mind. In this time of fast advancement, the mind has elevated affectability to the earth, where annoyance may incline the hatchling to postnatal neurodevelopmental issue. Generally speaking, supply of supplements during the bias and prebirth periods not just gives the fundamental structure squares to the mind, yet may likewise "program" the cerebrum through epigenetic components to present hazard or flexibility to neurological conditions sometime down the road. The modifiable idea of maternal nourishment during touchy periods conceivably offers open doors for intercession.[41–43]

6.3.2 NUTRACEUTICALS

6.3.2.1 ANTIOXIDANT VITAMIN SUPPLEMENTS (VITAMINS C AND E)

Antioxidant vitamin enhancements, for example, vitamin C, vitamin E (or tocopherol), and beta-carotene are regular types of nutraceuticals. A cross-sectional investigation found that vitamin E enhancements are prevalent in PD patients, while epidemiological examinations have indicated that devouring nourishments plentiful in vitamins C and E are related with a lower danger of creating PD. Antioxidant vitamins have a putative job in diminishing the oxidative harm in SN dopaminergic neurons in dynamic ailment. Nutrient C has been demonstrated in vitro to be a significant free-radical forager in the cytosol, while tocopherols go about as a significant lipid-solvent cell reinforcement to avoid lipid peroxidation in films. The two nutrients likewise act in a synergistic way whereby nutrient C can lessen oxidized nutrient E to reestablish its antioxidative capacity. In this manner, supplemental nutrients can be helpful in anticipation or in deferring movement of PD by decreasing oxidative pressure.

Vitamin D

Vitamin D has been found to weaken 6-hydroxydopamine 6-OHDA-actuated and 1-methyl-4-phenylpyridium ion (MPP+)-initiated neurodanger, while vitamin D receptor knockout mice show engine imperfection. Additionally, the degrees of vitamin D-restricting protein have been proposed as one of the biomarkers for PD. It has been discussed that nutrient D deficiency in PD patients is a consequence of decreased physical movement and presentation to daylight, instead of a causal factor in PD movement. The SN is one of the areas in the mind containing significant levels of vitamin D receptors and 1a-hydroxylase, the catalyst answerable for the organic enactment of vitamin D. Henceforth, nutrient D might be associated with various flagging pathways, and a few instruments might be answerable for the neuroprotective impacts of vitamin D.

Coenzyme Q10

Coenzyme Q10 (CoQ10 or ubiquinone) is a famous financially accessible dietary enhancement. It has been perceived as a neuroprotective operator in the aversion and treatment of PD. CoQ10 has been exhibited to anticipate the loss of dopaminergic neurons in methyl-4-phenyl-1,2,3, tetrahydropyridine (MPTP)-incited neurotoxicity, and Parkinsonism. CoQ10 is a fat-solvent and nutrient-like quinone discovered liberally in liver and the cerebrum. CoQ10 is especially significant to mitochondrial brokenness in light of its special electron-tolerating property, which enables it to connect mitochondrial complex I with different buildings. CoQ10 assumes a significant job in keeping up appropriate exchange of electrons in the electron transport chain of mitochondria and, in this way, in the creation of ATP also. Accordingly, CoQ10 protectively affects dopaminergic neurons in the SN. Furthermore, it is a powerful cancer prevention agent and can apply its cell reinforcement impact by diminishing the oxidized type of alpha-tocopherol, which is significant in the anticipation of lipid peroxidation.

Creatine

Creatine has additionally been researched for its conceivable job in the treatment and counteractive action of PD. Creatine is viewed as neuroprotective because of its capacity to counter ATP exhaustion by expanding intracellular phosphocreatine levels. Phosphocreatine is a key player in the support of ATP levels, which thus are significant in synaptic action and skeletal muscle capacities.

Natural Sources of L-DOPA

Until this point in time, common L-DOPA has been found in a few plants having a place with *Mucuna* genus, for example, *Mucuna pruriens* (velvet bean or mucuna, the seeds of which, in 1937, were found to contain L-DOPA), *Stizolobiumdeer ingianum*, and *Vicia faba* (wide bean, in which L-DOPA was identified in 1913). *M. pruriens* (called "atmagupta" in India) is a climbing legume medicine tropical districts that incorporate India and Central and South America. The plant has been reported in Ayurvedic medication to treat a neurological issue bearing side-effects like those of PD and up to 10% of the plant's volume is L-DOPA. In ongoing years, velvet bean seed concentrate has been utilized for the treatment of PD in India. Most in-vitro examinations on normal L-DOPA sources center around mucuna. In 2004, Manyam et al. showed that mucuna seed powder contained significant measures of two neuroprotective specialists, to be specific nicotine adenine dinucleotide (NADH) and CoQ10. The two specialists ensure neurons against 6-OHDA danger by checking the hindrance of mitochondrial complex I action. NADH is additionally known to build dopamine levels by means of the upregulation of tyrosinehydrolase. Mucuna seed powder has likewise been found to ensure neurons against plasmid DNA and genomic DNA harm brought about by a mix of L-DOPA and divalent copper ions. Mucuna seed powder secures neurons against this sort of harm by chelating the divalent copper particles present, keeping them from communicating with L-DOPA to create the free radicals that will harm DNA atoms.

Curcuminoids in Curry

Curcumin (1,7-bis[4-hydroxy-3-methoxyphenyl]-1,6-heptadiene-3,5-dione) is a polyphenolic flavonoid that comprises roughly 4% of turmeric, which has a long history of utilization in traditional Asian weight control plans and home-grown prescriptions. Curcumin is the head curcuminoid in turmeric. The other two curcuminoids are desmethoxy curcumin and bis-desmethoxy curcumin. The bioactive impacts of curcuminoids have regularly been credited to curcumin, as the curcumin substance of curcuminoids reaches up to 80%. Like other polyphenolic mixes, for example, caffeic corrosive, epigallocatechin gallate (EGCG), and resveratrol, curcumin is outstanding for its ground-breaking cancer prevention agent properties. Curcumin has likewise been found to show against inflammatory properties.

Baicalein

Baicalein is a flavonoid extracted from the base of *Scutellaria baicalensis*, at traditional Chinese herb usually known as Huang Qin. Baicalein has been demonstrated to be a powerful cancer prevention agent in rodent essential neurons. Another investigation in rodents likewise demonstrated enemy of inflammatory properties of baicalein in exploratory horrible cerebrum injury. Baicalein was seen as neuroprotective in a few test models of PD, including MPTP-actuated neurotoxicity and 6-OHDA-instigated neurotoxicity. It has additionally been appeared to hinder fibrillation and disaggregate the existing fibrils in the brain.

Stilbenes

Stilbenes are a class of cancer prevention agents having a similar concoction skeleton of a diaryl ethene, which is a hydrocarbon comprising of a *trans*/*cis*-ethene twofold bond substituted with a phenyl bunch on both carbon particles of the twofold bond. The name "stilbene" was gotten from the Greek word "stilbos," which signifies "sparkling." Many stilbenes and their derivates (stilbenoids) are normally present in plants (dietary organic products or herbs).

Resveratrol

The most generally examined stilbene is resveratrol (RES), a phyto-alexin found in plants, for example, grapes, peanuts, berries, and pines. RES is incorporated in these plants to check different ecological wounds, for example, UV light and contagious contamination. RES is accounted for to be one of the dynamic specialists in Itadori tea, which has been utilized as a customary prescription in China and Japan, basically for treating coronary illness and stroke. Epidemiological examinations detailing the backward relationship between moderate utilization of red wine and the occurrence of coronary illness have animated examinations on the cardioprotective movement of RES.

Phytoestrogens

It has been realized that the frequency of PD is lower in ladies than in men (utilizing age controls), showing a defensive impact of estrogen or its subordinates. The frequency of PD is likewise lower in premenopausal ladies than in postmenopausal women. The neuroprotective impacts of estrogen have been appeared in numerous investigations, including upregulation of Bcl-2 and cerebrum determined neurotropic factor. However, various reactions dishearten ladies from getting hormone substitution treatment. Phytoestrogens, acquired through either the eating routine or enhancements, give an option in contrast to customary hormone substitution treatment. Phytoestrogens are a gathering of substances that are found normally in plants and have a typical concoction structure like that of estradiol. Significant nourishment wellsprings of phytoestrogens incorporate soy items, nuts, and grains.

Ginsenoside Rg1

Ginsenosides are a class of particles extricated from a few types of ginseng. Ginseng has a long history in conventional Chinese medication, Indian natural drug, and the prescription of other Asian societies, and it is outstanding for its antiaging impacts. Rg1 is a ginsenoside detached from the base of Panaxginseng. It is one of the generally well-considered ginsenosides. In vivo, Rg1 can lessen 6-OHDA neurotoxicity, MPTP-incited neurotoxicity, and oxidative pressure. It can likewise stifle tumor expansion.

Genistein

Soy and peanuts are rich dietary wellsprings of the phyto-estrogen genistein, which has been seen as the essential flowing soy isoflavone. In fact, dietary soy is broadly utilized as an option in contrast to conventional hormonal substitution treatment. In 2007, an investigation was led by Azad Bakhtetal to find the impacts of dietary soy on postmenopausal ladies with metabolic disorder. Contrasted and typical subjects, the postmenopausal ladies had diminished plasma levels of malondialdehyde, an oxidative pressure marker. Various investigations in rodents have indicated that treatment with genistein separated from plant sources brings about comparable antioxidative impacts and hostile to apoptotic impacts.

Also, it has been recommended that genistein might have the option to direct movement of dopaminergic neurons since estradiol has been appeared to assume a job in guideline of the synapse in creature studies. A late examination testing the impacts of genistein treatment before intrastriatal 6-OHDA sores in rodents is in accordance with this speculation. It was discovered that genistein pretreatment weakened rotational conduct in rodents, a side-effect of parkinsonism.

Vitamins and Hyperhomocysteinemia

Various investigations have exhibited that treatment with L-DOPA in PD patients initiates elevated levels of homocysteine (HHcy). Studies show that HHcy is a considerable hazard factor for cardiovascular, cerebrovascular, and fringe vascular sicknesses just as subjective weakness and dementia.

Vitamin C, Hydrosoluble Fiber, and Pharmacokinetics

In spite of the fact that findings about the adequacy of the neuroprotective impacts of nutrient C were indecisive, nutrient C may improve the viability of L-DOPA. In a pharmacokinetic study, nutrient C was found to upgrade assimilation of L-DOPA in old patients with PD. Another investigation utilizing water-solvent fiber of *Plantago ovata* husk demonstrated that treatment of the plant with L-DOPA/carbi-DOPA benefits PD patients by soothing obstruction and improving the L-DOPA profile.

Flavonoids

Flavonoids are a gathering of poliphenolic aggravates that are regular in the every-day human eating routine. They are found in many plants, including organic products, vegetables, and a few sorts of normal beverages, for example, tea, cocoa, and wine. Flavonoids and their metabolites adjust a few neurological procedures as appeared by thinks about in which an association with neuronal–glial flagging pathways engaged with neuronal endurance and capacity was watched. What's more, flavonoids incite changes in cerebral bloodstream, upregulate cancer prevention agent compounds and proteins associated with synaptic versatility and neuronal fix and restrain neuropathological forms in mind districts regularly engaged with AD pathogenesis.

Curcumin (Curcuma longa L.)

Curcuma longa Linn. is an enduring herb having a place with the family Zingiberaceae. Turmeric is gotten from the rhizome of the curcumin plant and is ordinarily utilized in India as an enhancing and shading specialist in nourishment. Curcumin has been utilized for a considerable length of time in Ayurvedic prescription, and it is the most methodically read plant for its advantages in various infections. Significant constituents of *C. longa* are curcumin, demethoxycurcumin, and bis-demethoxy curcumin by and large known as curcuminoids. The capability of curcumin in treating neurodegenerative sicknesses has been very much archived.

Ashwagandha (Withania somnifera)

Withania somnifera, a plant having a place with the Solanaceae family, otherwise called Indian Ginseng or Ashwagandha, is alluded to as "Medhyarasayan" (nootropic herb) in Ayurvedic drug. At the same time, Ashwagandha has been customarily used to treat different wellbeing conditions, including mental pressure, tension, depression, and memory misfortune, notwithstanding its general wellbeing utility as an incredible cell reinforcement, neuroprotective, and mitigating drug.

Brahmi (Bacopa monnieri)

Bacopa monnieri, having a place with the Scrophulariaceae family and generally called "Brahmi," is known for its rejuvenating and nootropic exercises in Ayurvedic medicine as it fortifies memory and insight. The alcoholic concentrate of *Bacopa monnieri* improved securing, combination, and maintenance of memory in a foot stun roused brilliance separation test and a conditioned-shirking test in rodents. Bacosides A and B (a blend of two saponins) might be liable for its advancing impact on learning and memory. Organization of bacosides (200 mg/kg) for a quarter of a year in matured rodents applied a defensive impact against age-related modifications in the neurotransmission framework, conducted ideal models, hippocampal neuronal misfortune, and oxidative pressure markers.

Almonds (Prunus dulcis/Amygdalus L.) and Walnuts (Juglans regia L.)

In Ayurveda, an eating regimen containing a plenitude of nuts which is viewed as a "Saatvic" diet (a yogic eating routine that advances imperativeness, vitality, energy, and wellbeing). Nuts are profoundly nutritious, plentiful in fats, protein, nutrients, starches, and contain a variety of phytochemicals, including carotenoids, phenolic acids, polyphenols, flavonoids, ligans, phytosterols, and the sky is the limit from there. These phytochemicals found in nuts are very well studied for their antioxidant, mitigating, antitumor, antimicrobial, chemopreventive, hypocholesterolemic, neuroprotective, and nootropic impacts. Albeit, all nuts are known for their medical advantages, almonds and pecans specifically have been the most broadly read for their nootropic impacts. Almonds have been appeared to apply neuroprotective and nootropic effects in numerous exploratory models.

6.4 CONCLUSION

Nutraceuticals have demonstrated their medical advantages and malady anticipation ability, which ought to be taken by their satisfactory prescribed admission. The present situation of self-drug nutraceuticals assume significant job in restorative improvement. Be that as it may, their prosperity relies upon keeping up on their quality, virtue, security, and viability. Nature has skilled human with an abundance of assets that are basic in keeping up our wellbeing. The availability of bioactive supplements from restorative herbs, natural products, vegetables, and nuts has extraordinary potential for avoiding and relieving numerous wellbeing issues, including neurodegeneration. Gathering proof demonstrates the importance of the pretended by phytonutrients in deferring neurodegeneration and improving cognizance. The curative impacts of these bioactive nutraceuticals might be ascribed to their cancer prevention agent, mitigating, hypocholesterolemic, neuroprotective, and nootropic activities went for unmistakable subatomic targets engaged with protein misfolding and the outcomes usually saw in neurodegenerative ailments. Albeit few investigations have been directed in people, late advancement in exploratory research including plant items is rising and may open up new roads for elective, moderate treatment for avoiding, and restoring neurodegenerative maladies, which speak to a developing general wellbeing emergency.

KEYWORDS

- **nutraceutical**
- **neurodisorder**
- **natural product**
- **nutrition**
- **treatment**
- **vitamins**

REFERENCES

1. Sarin, R.; et al. Nutraceuticals: A Review. *Int. Res. J. pharm.* **2012,** *3* (4), 95–99.
2. Chauhan, B.; Kumar, G.; Kalam, N.; Ansari, S. H. Current Concepts and Prospects of Herbal Nutraceutical: A Review. *J. Adv. Pharm. Technol.* **2013,** *4* (1), 4–8.
3. Kalra, E. K. Nutraceutical—Definition and Introduction. *AAPS Pharm. Sci.* **2003,** *5* (3), 27–28.
4. Sahu, M.; Sapkale, P. A. Review on Palladium Catalyzed Coupling Reactions. *Int. J. Pharm. Chem. Sci.* **2013,** *2* (3), 1159–1170.
5. Pandey, M.; Verma, R. K.; Saraf, S. A. Nutraceuticals New Era of Medicine and Health. *Asian J. Pharm. Clin. Res.* **2010,** *3* (1), 1–5..
6. Chauhan, B.; Kumar, G.; Kalam, N.; Ansari, S. H. Current Concepts and Prospects of Herbal Nutraceuticals: A Review. *J. Adv. Pharm. Technol. Res.* **2013,** *4* (10), 4–8.
7. Clare, L.; et al. Biological Basis for the Benefit of Nutraceutical Supplementation in Arthritis. *Drug Discov. Today* **2004,** *9* (4), 165–172.
8. Emam, M. A.; et al. *J. Nutr. Metab.* **2012**.
9. Gursevak, S.; et al. Functional Foods and Nutraceuticals in the Management of Obesity. *Nutr. Food Sci.* **2005,** *35*, 344–352.
10. Heredia, F. P.; et al. Functional Foods and Nutraceuticals as Therapeutic Tools for the Treatment of Diet Related Diseases. *Can. J. Physiol. Pharmacol.* **2013,** *91* (6), 387–396.
11. Ahmad, M.; Saleem, S.; Ahmad, A. S.; Yousuf, S.; Ansari, M. A.; Khan, M. B.; et al. *Ginkgo biloba* Affords Dose-Dependent Protection against 6-Hydroxydopamine-Induced Parkinsonism in Rats: Neurobehavioural, Neurochemical and Immunohistochemical Evidences. *J. Neurochem.* **2005,** *93* (1), 94–104.
12. Albani, D.; Polito, L.; Batelli, S.; De Mauro, S.; Fracasso, C.; Martelli, G.; et al. The SIRT1 Activator Resveratrol Protects SK-N-BE Cells from Oxidative Stress and against Toxicity Caused by alpha-synuclein or amyloid-beta (1–42) Peptide. *J. Neurochem.* **2009,** *110* (5), 1445–1456.
13. Ali, S.; Shultz, J. L.; Haq, I. High Performance Microbiological Transformation of l-tyrosine to l-dopa by *Yarrowia lipolytica* NRRL-143. *BMC Biotechnol.* **2007,** *7*, 50.
14. Anderson, D. W.; Bradbury, K. A.; Schneider, J. S. Broad Neuroprotective Profile of Nicotinamide in Different Mouse Models of MPTP-Induced Parkinsonism. *Eur. J. Neurosci.* **2008,** *28* (3), 610–617.
15. Baluchnejadmojarad, T.; Roghani, M.; Nadoushan, M. R.; Bagheri, M. Neuroprotective Effect of Genistein in 6-Hydroxy-dopamine Hemi-Parkinsonian Rat Model. *Phytother. Res.* **2009,** *23* (1), 132–135.
16. Beal, M. F.; Matthews, R. T.; Tieleman, A.; Shults, C. W. Coenzyme Q10 Attenuates the 1-Methyl-4-phenyl-1,2,3, Tetrahydro-pyridine (MPTP) Induced Loss of Striatal Dopamine and Dopaminergic Axons in Aged Mice. *Brain Res.* **1998,** *783* (1), 109–114.
17. Bender, A.; Koch, W.; Elstner, M.; Schombacher, Y.; Bender, J.; Moeschl, M.; et al. Creatine Supplementation in Parkinson Disease: A Placebo-Controlled Randomized Pilot Trial. *Neurology* **2006,** *67* (7), 1262–1264.

18. Jiang, M.; Porat-Shliom, Y.; Pei, Z.; Cheng, Y.; Xiang, L.; Sommers, K.; et al. Baicalein Reduces E46K Alpha-synuclein Aggregation *In Vitro* and Protects Cells against E46K Alpha-synuclein Toxicity in Cell Models of Familiar Parkinsonism. *J. Neurochem.* **2010,** *114* (2), 419–429.
19. Johnson, S. Micronutrient Accumulation and Depletion in Schizophrenia, Epilepsy, Autism and Parkinson's Disease? *Med. Hypoth.* **2001,** *56* (5), 641–645.
20. Kang, X.; Chen, J.; Xu, Z.; Li, H.; Wang, B. Protective Effects of *Ginkgo biloba* Extract on Paraquat-Induced Apoptosis of PC12 Cells. *Toxicol. Vitro* **2007,** *21* (6), 1003–1009.
21. Katzenschlager, R.; Evans, A.; Manson, A.; Patsalos, P. N.; Ratnaraj, N.; Watt, H.; et al. *Mucuna pruriens* in Parkinson's Disease: A Double Blind Clinical and Pharmacological Study. *J. Neurol. Neurosurg. Psychiatry* **2004,** *75* (12), 1672–1677.
22. Kaul, S.; Anantharam, V.; Yang, Y.; Choi, C. J.; Kanthasamy, A.; Kanthasamy, A. G. Tyrosine Phosphorylation Regulates the Proteolytic Activation of Protein Kinase Cd in Dopaminergic Neuronal Cells. *J. Biol. Chem.* **2005,** *280* (31), 28721–28730.
23. Khan, M. M.; Ahmad, A.; Ishrat, T.; Khan, M. B.; Hoda, M. N.; Khuwaja, G.; et al. Resveratrol Attenuates 6-Hydroxy-Dopamine-Induced Oxidative Damage and Dopamine Depletion in Rat Model of Parkinson's Disease. *Brain Res.* **2010,** *1328*, 139–151.
24. Pennesi, C. M.; Klein, L. C. Effectiveness of the Gluten-Free, Casein-Free Diet for Children Diagnosed with Autism Spectrum Disorder: Based on Parental Report. *Nutr. Neurosci.* **2012,** *15* (2), 85–91.
25. Ploeger, A.; Galis, F. Evolutionary Approaches to Autism—An Overview and Integration. *MJM* **2011,** *13* (2), 38–43.
26. Rapin, I. The Autistic Spectrum Disorders. *N. Engl. J. Med.* **2002,** *347* (5), 302–304.
27. Richardson, A. J. Long-Chained Polyunsaturated Fatty Acids in Childhood Developmental and Psychiatric Disorders. *Lipids* **2004,** *39*, 1215–1222.
28. Schmidt, R. J.; Hansen, Rl; Hartiala, J.; Allayee, H.; Schmidt, L. C.; Tancredi, D. J.; Tassone, F.; Hertz-Picciotto, I. Prenatal Vitamins, One-Carbon Metabolism Gene Variants, and Risk for Autism. *Epidemiology* **2011,** *22*, 476–485.
29. Tani, Y.; Fernell, E.; Watanabe, Y.; Kanai, T.; Langstrom, B. Decrease in 6R-5,6,7,8-tetrahydrobiopterin Content in Cerebrospinal Fluid of Autistic Patients. *Neurosci. Lett.* **1994,** *181*, 169–172.
30. Tonge, B.; Brereton, A. Autism Spectrum Disorder. *Aust. Fam. Physician* **2011,** *40* (9), 672–677.
31. Stough, C.; Lloyd, J.; Clarke, J.; Downey, L. A.; Hutchison, C. W.; Rodgers, T.; Nathan, P. J. The Chronic Effects of an Extract of *Bacopa monniera* (Brahmi) on Cognitive Function in Healthy Human Subjects. *Psychopharmacology (Berl.)* **2001,** *156*, 481–484.
32. Stough, C.; Scholey, A.; Cropley, V.; Wesnes, K.; Zangara, A.; Pase, M.; Savage, K.; Nolidin, K.; Lomas, J.; Downey, L. Examining the Cognitive Effects of a Special Extract of *Bacopa monniera* (CDRI08: Keenmnd): A Review of Ten Years of Research at Swinburne University. *J. Pharm. Pharm. Sci.* **2013,** *16*, 254–258.
33. Suh, S. J.; Koo, B. S.; Jin, U. H.; Hwang, M. J.; Lee, I. S.; Kim, C. H. Pharmacological Characterization of Orally Active Cholinesterase Inhibitory Activity of Prunus persica L. Batsch in RATS. *J. Mol. Neurosci.* **2006,** *29*, 101–107.

34. Tohda, C.; Joyashiki, E. Sominone Enhances Neurite Outgrowth and Spatial Memory Mediated by the Neurotrophic Factor Receptor, RET. *Br. J. Pharmacol.* **2009,** *157*, 1427–1440.
35. Tohda, C.; Kuboyama, T.; Komatsu, K. Dendrite Extension by Methanol Extract of Ashwagandha (roots of *Withania somnifera*) in SK-N-SH Cells. *NeuroReport* **2000,** *11*, 1981–1985.
36. Vollala, V. R.; Upadhya, S.; Nayak, S. Learning and Memory-Enhancing Effect of *Bacopa monniera* in Neonatal Rats. *Bratisl. Lek. Listy* **2011,** *112*, 663–669.
37. House, J. S.; Mendez, M.; Maguire, R. L.; Gonzalez-Nahm, S.; Huang, Z.; Daniels, J.; Murphy, S. K.; Fuemmeler, B. F.; Wright, F. A.; Hoyo, C. Periconceptional Maternal Mediterranean Diet Is Associated with Favorable Offspring Behaviors and Altered CpG Methylation of Imprinted Genes. *Front. Cell Dev. Biol.* **2018,** *6*, 107.
38. Schmidt, R. J.; Ozonoff, S.; Hansen, R.; Hartiala, J.; Allayee, H.; Schmidt, L.; Tassone, F.; Hertz-Picciotto, I. (Old Record 6424) Maternal Periconceptional Folic Acid Intake and Risk for Developmental Delay and Autism Spectrum Disorder: A Case-Control Study. *Am. J. Epidemiol.* **2012,** *175*, S126.
39. Li, Y. M.; Ou, J. J.; Liu, L.; Zhang, D.; Zhao, J. P.; Tang, S. Y. Association between Maternal Obesity and Autism Spectrum Disorder in Offspring: A Meta-analysis. *J. Autism Dev. Disord.* **2016,** *46*, 95–102.
40. Ueland, P. M.; Hustad, S.; Schneede, J.; Refsum, H.; Vollset, S. E. Biological and Clinical Implications of the MTHFR C677T Polymorphism. *Trends Pharmacol. Sci.* **2001,** *22*, 195–201.
41. Stamm, R.; Houghton, L. Nutrient Intake Values for Folate during Pregnancy and Lactation Vary Widely around the World. *Nutrients* **2013,** *5*, 3920–3947.
42. Schmitz, G.; Ecker, J. The Opposing Effects of n-3 and n-6 Fatty Acids. *Prog. Lipid Res.* **2008,** *47*, 147–155.
43. Patterson, E.; Wall, R.; Fitzgerald, G. F.; Ross, R. P.; Stanton, C. Health Implications of High Dietary Omega-6 Polyunsaturated Fatty Acids. *J. Nutr. Metab.* **2012,** *2012*, 539426.

CHAPTER 7

1,3-Diphenyl-2-Propene-1-One-Based Natural Product Antidiabetic Molecules as Inhibitors of Protein Tyrosine Phosphatase-1B (PTP-1B)

DEBARSHI KAR MAHAPATRA[1*], SANJAY KUMAR BHARTI[2], and VIVEK ASATI[3]

[1]*Department of Pharmaceutical Chemistry, Dadasaheb Balpande College of Pharmacy, Nagpur 440037, India*

[2]*Institute of Pharmaceutical Sciences, Guru Ghasidas Vishwavidyalaya (A Central University), Bilaspur 495009, India*

[3]*Department of Pharmaceutical Chemistry, NRI Institute of Pharmacy, Bhopal 462021, India*

**Corresponding author. E-mail: mahapatradebarshi@gmail.com*

ABSTRACT

The chapter has comprehensively focused on some very unknown natural chalcone compounds (kuwanon J, kuwanon R, kuwanon V, isoliquiritigenin, xanthoangelol, xanthoangelol D, xanthoangelol E, xanthoangelol F, xanthoangelol K, 4-hydroxyderricin, 5,4'-dihydroxy-6,7-furanbavachalcone, licochalcone A, licochalcone B, licochalcone C, licochalcone D, licochalcone E, echinatin, laxichalcone, broussochalcone, macdentichalcone, (2*E*)-1-(5,7-dihydroxy-2,2-dimethyl-2*H*-benzopyran-8-yl)-3-phenyl-2-propen-1-one, (2*E*)-1-(5,7-dihydroxy-2,2,6-trimethyl-2*H*-benzopyran-8-yl)-3-(4-methoxyphenyl)-2-propen-1-one, and abyssinone-VI-4-O-methyl ether) having tremendous potential to exhibit antidiabetic activity by selectively modulating the promising therapeutic target protein tyrosine phosphatase 1B (PTP-1B) that will prevent the degradation of insulin. In modern days, these natural product chalcone-based PTP-1B inhibitors are not under clinical use and they have not received any such attention in modern-day medicine as they are not explored clinically in terms of toxicological profiles to develop a suitable formulation. In the near future, it is expected that these chalcone-based PTP-1B inhibitors will open new avenues of diabetotherapeutics.

7.1 INTRODUCTION

Diabetes mellitus (DM) is exclusively characterized by enhanced plasma sugar levels along with several groups of heterogeneous disorders such as alteration in the metabolism of proteins, carbohydrates, and lipids.[1] This constant hyperglycemic circumstance leads to enhanced risks of vascular complications and directly affects the blood vessels, eyes, kidneys, heart, and nerves.[2] A patient suffering from DM experiences hepatic gluconeogenesis, reduced uptake of blood glucose by tissues, and impaired insulin secretion which concurrently results in precipitation of symptoms like excessive hunger, random plasma blood sugar level of >200 mg/dL, glucose in the urine, weight loss, and excessive thirst.[3–4]

In two discrete phases (fasting blood glucose concentration and postprandial blood glucose concentration), the secretion (magnitude) of insulin takes place from the β-cells of the pancreas.[5] First, a speedy release of insulin takes place just after the meal (due to rapid augmentation of glucose levels) which is pursued by a sustained phase of circulating concentrations of insulin. In DM, intrinsic problems such as ineffective hyperglycemic and hypoglycemic phases due to fluctuation of two phases of insulin are perceived.[6] In type-I DM, insulin deficiency leads to failure in the conversion of sugar into its storage form and utilization, whereas failure of proper utilization of secreted insulin is the chief characteristic of type-II DM. In type-II DM, reduced adipose cells and muscle sensitivity toward insulin are the most prominent features.[7]

In the pharmacotherapeutic point of view, insulin sensitizers are the best compounds for the successful treatment of the hyperglycemic conditions that will amplify the muscle and adipose tissue's sensitivity to insulin.[8] In the modern era, glitazones and sulfonylureas are not much effective in the management of hyperglycemic episodes and therefore the need for effective inhibitors is a major challenge.[9] For diabetotherapy, protein tyrosine phosphatase-1B (PTP-1B) inhibitors and dipeptidyl Peptidase-4 (DPP-4) inhibitors are the upcoming preferred options as these compounds prevent the degradation of insulin and prolong the action.[10]

7.2 PROTEIN TYROSINE PHOSPHATASE-1B (PTP-1B)

For mediating various intracellular functions like insulin action, metabolic activities, etc., the enzyme class of protein kinases and phosphatases play a dominant role by mediating the dephosphorylation and phosphorylation reactions.[11] A non-receptor, intercellular PTP, known as PTP-1B is present in the cytoplasm of endoplasmic reticulum and is considered a classic target for high insulin uptake tissues like adipose tissue, muscles, and liver.[12] PTP-1B is a negative regulator of the insulin in the leptin signaling pathway and plays a key role in the management of type-II DM.[13] The insulin signaling mediation occurs through insulin receptor (IR) activation by autophosphorylation on tyrosine residues. In the process of dephosphorylation of IR, a number of PTPs like SH2-domain-containing phosphotyrosine phosphatase (SHP2), receptor protein tyrosine phosphatase (rPTP-a), PTP-1B, and leukocyte antigen-related (LAR) tyrosine phosphatase play major functions.[14] The mode of insulin signaling down regulation takes place by dephosphorylation of IR, IR substrate-1 (IRS-1), and IR substrate-2 (IRS-2).[15]

The kinase enzyme, phosphoinositide 3-kinases (PI3K) downstream the metabolic signaling by phosphorylating the substrate PI to phosphatidylinositol biphosphate (PIP2), thereby activates the protein 3-phosphoinositide-dependent protein kinase-1 (PDK1).[16] This cascade activates protein kinase-B (PKB) which is a sole component in the enrichment of glucose uptake by stimulating insulin-dependent GLUT4 translocation (Fig. 7.1).[17]

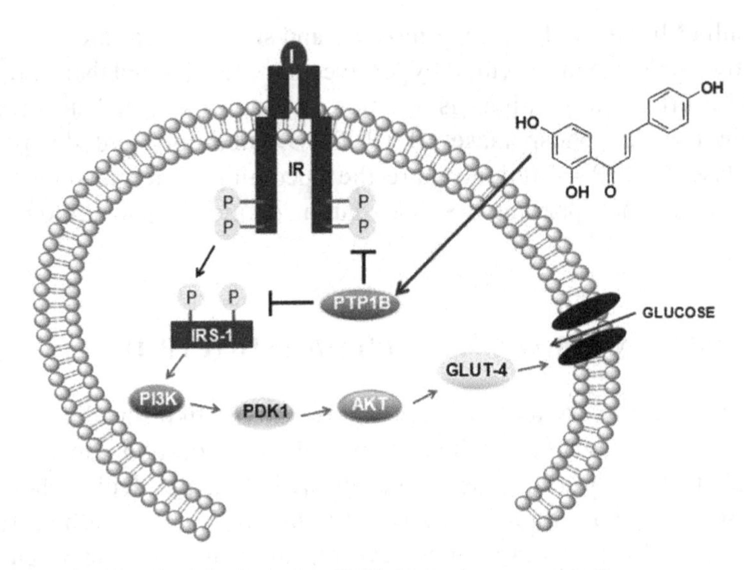

FIGURE 7.1 The physiological role of PTP-1B in glucose metabolism.

The catalysis of PTP-based reactions take place via following two mechanistic steps; the first step involves a nucleophilic attack on the substrate phosphate by the –S– atom of Cys residue (situated in the thiolate side chain) where a coupled protonated tyrosyl-leaving group of side chain (Asp 181) acted as conserved acidic residue, thereby formed an intermediate for cysteinyl-phosphate catalysis. Step two is mediated by the residue Gln 262, coordinated through H_2O and the conserved acidic residue that functions as a base component that causes catalytic intermediate hydrolysis and release of a phosphate molecule (Fig. 7.2).[18]

7.3 ROLE OF PTP-1B IN DIABETES MELLITUS

Present-day studies have indicated that PTP1B knockout animals demonstrated improved glucose tolerance, reduction in diet-induced obesity, and enhanced sensitivity of cells toward insulin.[19] Similarly, the administration of PTP-1B antisense oligonucleotides showed improved insulin sensitivity and normalized the plasma glucose levels as a result of reduced enzyme expression.[20] Clinical researches have indicated the role of PTP-1B inhibitors (*ertiprotafib*; discontinued from clinical trials due to lack of efficacy and *trodusquemine*; presently under clinical trials) in the dephosphorylation of IR and down regulation of insulin signaling pathway.[21] The tyrosine mimetic comprising of negatively charged functionalized components like malonates, cinnamates, phosphonates, and carboxylates have been recognized as PTP-1B inhibitors with distinct advantages.[22] These above collective evidence from genetic, pharmaceutical, physiological, and biochemical backgrounds have addressed toward the perspectives of PTP-1B inhibitors as the latest candidates in the management of hyperglycemic episodes and related obesity. This exciting antidiabetic target (specifically type-II DM) will be of immense significance toward the development of potent low-molecular-weight inhibitors.[23]

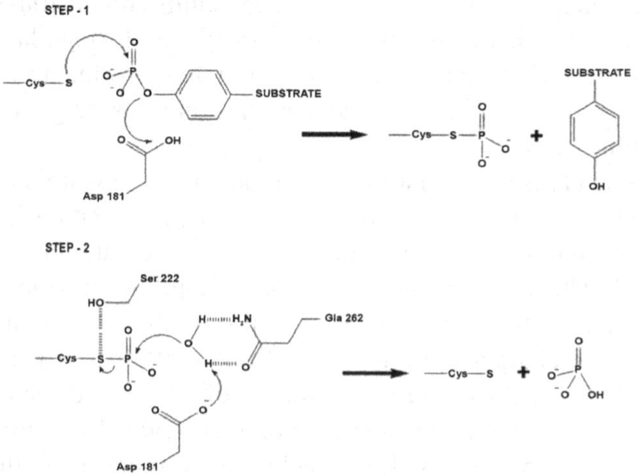

STEP - 1

STEP - 2

FIGURE 7.2 The process and biochemical pathway of PTP-mediated catalysis in human body.

However, several PTP-1B inhibitors are available and reported widely but poor pharmacokinetic properties and low selectivity remained a challenging issue.[24] For meeting the need for better PTP-1B inhibitors in the current scenario, natural chalcones have been recently been identified in the management of major diabetic complications with better selectivity and also without any pharmacokinetic compromise. In modern days, these natural product chalcone-based inhibitors are not under clinical use and they have not received any such attention as they are not explored clinically in terms of toxicological profiles to develop a suitable formulation.

7.4 NATURAL PRODUCT CHALCONES

Natural products have been found to express tremendous antihyperglycemic activity.[25] Flavonoids (comprising of heat stable polyphenols) are the heterogeneous group of compounds that remained the most privileged class showing multifarious pharmacological activities (provide effective hypoglycemic control, reduce the risk of tumors formation, etc.).[26] There are >5000 polyphenolic compounds that existed all among the plant kingdom for billions of years which offer numerous health benefits (modulatory, mimetic, and protective properties).[27] These beneficial natural products exist in fruits and vegetables and are broadly classified into six important classes: flavan-3-ols (e.g., theaflavin, catechin, and gallic esters of catechins and theaflavins, etc.), isoflavones (e.g., daidzein, genistein, etc.), flavonols (e.g., kaempferol, quercetin, etc.), flavanones (e.g., naringenin, hesperidin, etc.), flavones (e.g., luteolin, apigenin, etc.), and anthocyanidins (e.g., cyaniding, pelargonidin, etc.).[28]

Chalcone **(1)** is a precursor of flavonoid and isoflavonoid originated from Mother Nature.[29] It is also known as 1,3-diphenyl-2-propene-1-one that comprises of an open-chain intermediate in the aurone synthesis.[30] The benzylideneacetophenone component comprising of two aromatic components linked by α, β-unsaturated carbonyl bridge function is the characteristic identity of this scaffold.[31] It is one of the most privileged scaffolds due to feasibility of productive computational studies,[32] synthesized at academic laboratory scale (first fabricated as chromophoric products by Kostanecki and Tambor in 19th century from benzaldehyde and acetophenone),[33] therapeutic targets modulating ability (actions against protozoal infections, bacterial infections, sleep disorder, trypanosomiasis, anxiety, gout, malaria, epilepsy, hypertension, reduced immune response, spasm, diabetes, tuberculosis, leishmaniasis, ulcer, fungal infections, cancer, thrombosis, reactive oxygen species, inflammation, HIV, metastasis, etc.),[34–36] simplicity in the overall product chemistry, structural elucidation of natural products (chromanochromane, tannin, flavanone, flavonoid, etc.),[37] development at industrial or commercial scales (high-yield methods such as one-pot reactions, Claisen-Schmidt reaction, Microwave-assisted reactions, Suzuki-Miyaura reaction, Direct crossed-coupling reaction, Friedel-Crafts reaction, Julia-Kocienski reaction, Sonogashira isomerization coupling, Solvent-free reactions, solid acid catalyst-mediated reactions, Carbonylated Heck coupling reaction, etc.),[38] hybridization with multiple other scaffolds (benzodiazepine, pyrazoline, thiadiazole, pyrimidine, benzoxazepine, isoxazole, pyrazole, benzothiazepine, etc.),[39] various non-pharmacological applications (chromophores in dying industries, scintillator, catalyst in specific reactions, polymerization agents in product development, fluorescent polymeric agents, sweeteners in confectionaries, chemosensor for detection, fluorescent whitening agent, insecticides, etc.),[40–41] and 10 replaceable hydrogen atoms for producing multiple active compounds.[42]

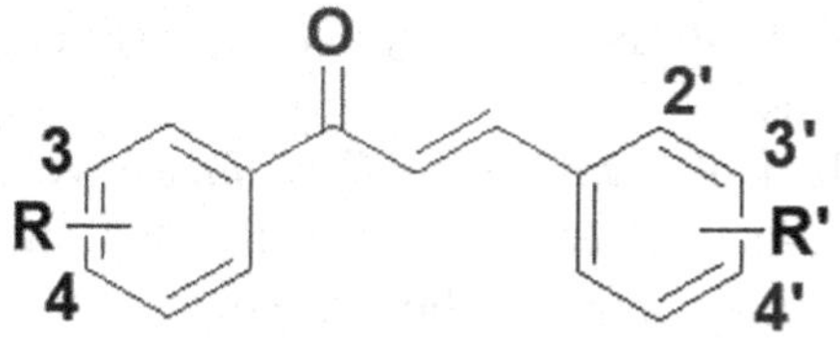

O
3 2' 3'
R R'
4 4'

(1)

7.5 NATURAL CHALCONES AS PTP-1B INHIBITORS

A handful of natural chalcone molecules have been reported to express potent PTP-1B inhibition activity (Fig. 7.3). From the leaf extract of *Broussonetia papyrifa*, broussochalcone **(2)** was isolated and screened for PTP-1B inhibitory potential with an IC_{50} value of 21.5 μm. The two –OH groups situated at both the rings of the compound are responsible for the inhibitory activity by interacting with the active sites of the enzyme. It was predicted that the inhibitory activity increases with an increase in the number of –OH groups in the pharmacophore.[43]

Kuwanon J **(3)**, kuwanon R **(4)**, and kuwanon V **(5)**, the methylcyclohexene substituted derived Diels-Alder type chalcones, isolated from *Morus bombycis* have demonstrated tremendous PTP-1B inhibitory activity (IC_{50} values in the range of 2.7–13.8 μm) in a mixed-type mechanism. The presence of the –OH group provides excellent penetration of the molecule into the active site of the therapeutic target and provides an effectual hydrogen bonding interaction with the active site loop of

PTP-1B. It is stated that with an increase in the number of –OH groups in Diels–Alder type compounds, the pharmacological potential augments simultaneously. The molecule **(4)** comprising of 7 –OH groups produced a dose-independent inhibition in comparison to compound **(5)** which has 6 –OH groups. The compound **(3)** has one more –OH group at second carbon atom which enhances the potency by three-folds as compared to compound **(5)**. Compound **(3)** has better pharmacological efficacy and potency than compound **(4)** and compound **(5)**.[44]

The coumarin-based chalcones; (2*E*)-1-(5,7-dihydroxy-2,2-dimethyl-2*H*-benzopyran-8-yl)-3-phenyl-2-propen-1-one **(6)**, (2*E*)-1-(5,7-dihydroxy-2,2,6-trimethyl-2*H*-benzopyran-8-yl)-3-(4-methoxyphenyl)-2-propen-1-one **(7)**, and laxichalcone **(8)** have presented noteworthy *in vitro* PTP-1B inhibitory activity with IC_{50} values in micro-molar concentrations.[45]

From the root bark of *Erythrina mildbraedii*, a novel chalcone molecule abyssinone-VI-4-O-methyl ether **(9)** was isolated and evaluated for *in vitro* PTP-1B inhibitory activity where a remarkable inhibition was seen with an IC_{50} value of 14.8 μm.[46]

A polycyclic dimeric chalcone comprising of quinonoid moiety isolated from *Macaranga denticulata*, macdentichalcone **(10)** along with its monomeric biosynthetic precursor 1-(5,7-dihydroxy-2,2,6-trimethyl-2*H*-1-benzopyran-8-yl)-3-phenyl-2-propen-1-one **(11)**. On *in vitro* PTP-1B inhibitory screening, both the compounds exhibited impressive pharmacotherapeutic activity with IC_{50} values of 21 μm and 22 μm, respectively.[47]

From the leaves of *Glycyrrhiza inflate*, isoliquiritigenin **(12)**, echinatin **(13)**, licochalcone A **(14)**, licochalcone C **(15)**, licochalcone E **(16)**, licochalcone B **(17)**, and licochalcone D **(18)** were isolated and reported to be promising candidates for the inhibition of PTP-1B in micromolar concentration.[48]

Isoliquiritigenin **(12)** was investigated to inhibit the antidiabetic target PTP-1B by preventing the phosphorylation of IR/PI3K/AKT and also inhibiting the oxidation of PTP-1B under insulin-induced adipogenesis stages and insulin-induced adipocyte differentiation of 3T3-L1 cells.[49]

The well-known natural chalcones; xanthoangelol K **(19)**, xanthoangelol **(20)**, xanthoangelol F **(21)**, 4-hydroxyderricin **(22)**, xanthoangelol D

(23), and xanthoangelol E **(24)** displayed amazing *in vitro* PTP-1B inhibition with IC_{50} range 0.82–3.98 μg/mL in a competitive manner. Docking studies have revealed that the ring-B of the natural chalcones anchors in the pocket of the antidiabetic target through hydrogen bonds (Arg 47 and Asp 48) and p-p interactions (Phe 182).[50]

Chalcone, (*E*)-1-(2,4-dihydroxy-3-(3-methylbut-2-en-1-yl) phenyl)-3-(2,4-dihydroxyphenyl)prop-2-en-1-one **(25)** was isolated from the root bark of *Morus alba* L. along with 21 phenolic compounds. The chalcone exhibited a notable PTP-1B inhibition with an IC_{50} value of 31.61 μm in a noncompetitive manner.[51]

A novel chalcone 5,4'-dihydroxy-6,7-furanbavachalcone **(26)** was isolated along with other polyphenols (isobavachalcone, bavachin, psoralenol, corylifol E, and corylifol A) from the seeds of *Psoralea corylifolia* (known as Bu-Gu-Zhi in traditional Chinese medicine). The novel chalcone expressed *in vitro* PTP-1B inhibition with an IC_{50} value of 14.3 μm.[52]

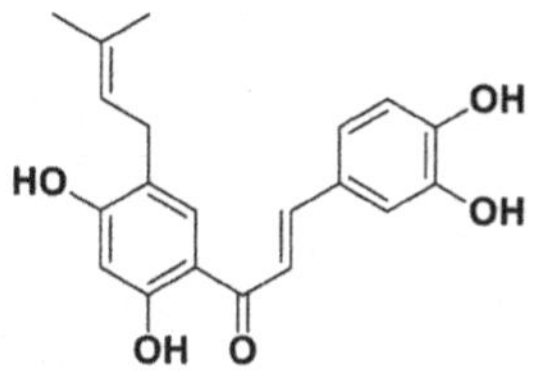

(2)

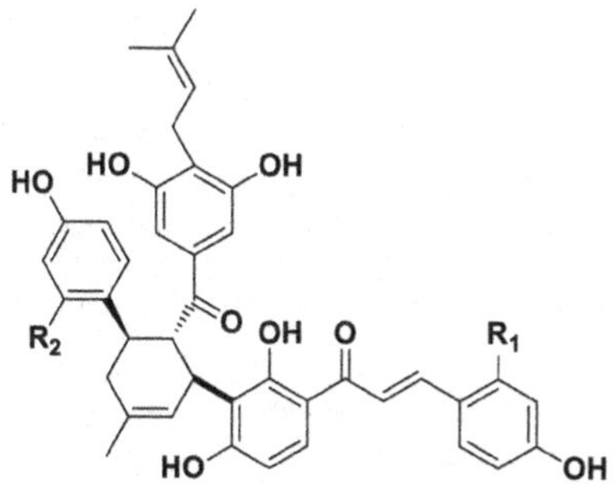

(3) R_1 = OH, R_2 = OH; **(4)** R_1 = H, R_2 = OH; **(5)** R_1 = H, R_2 = H

FIGURE 7.3 *(Continued)*

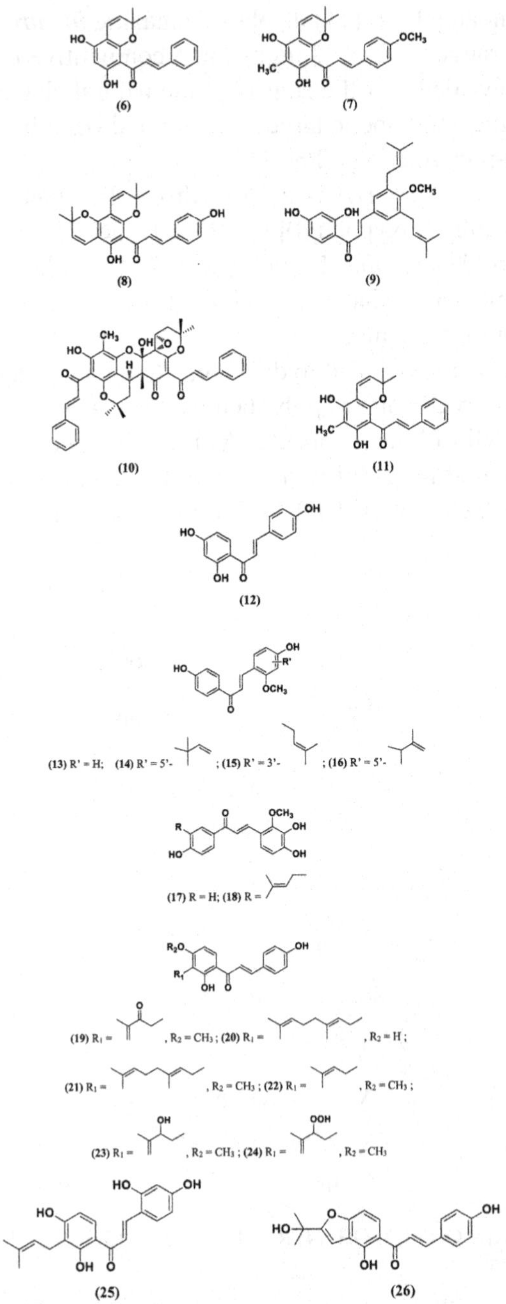

HO OH O (6)

HO OCH$_3$ H$_3$C OH O (7)

OH OH O (8)

OCH$_3$ HO OH O (9)

CH$_3$ HO OH O H O O O (10)

HO H$_3$C OH O (11)

OH HO OH O (12)

OH R' HO OCH$_3$ O

(13) R' = H; (14) R' = 5'- ; (15) R' = 3'- ; (16) R' = 5'-

O OCH$_3$ R OH HO OH

(17) R = H; (18) R =

R$_2$O OH R$_1$ OH O

(19) R$_1$ = , R$_2$ = CH$_3$; (20) R$_1$ = , R$_2$ = H ;

(21) R$_1$ = , R$_2$ = CH$_3$; (22) R$_1$ = , R$_2$ = CH$_3$;

(23) R$_1$ = OH , R$_2$ = CH$_3$; (24) R$_1$ = OOH , R$_2$ = CH$_3$

HO OH HO OH O (25)

OH HO OH O (26)

FIGURE 7.3 List of reported natural chalcone-based PTP-1B inhibitors.

7.6 CONCLUSION

The chapter has comprehensively focused on some very unknown natural chalcone compounds (kuwanon J, kuwanon R, kuwanon V, isoliquiritigenin, xanthoangelol, xanthoangelol D, xanthoangelol E, xanthoangelol F, xanthoangelol K, 4-hydroxyderricin, 5,4'-dihydroxy-6,7-furanbavachalcone, licochalcone A, licochalcone B, licochalcone C, licochalcone D, licochalcone E, echinatin, laxichalcone, broussochalcone, macdentichalcone, (2*E*)-1-(5,7-dihydroxy-2,2-dimethyl-2*H*-benzopyran-8-yl)-3-phenyl-2-propen-1-one, (2*E*)-1-(5,7-dihydroxy-2,2,6-trimethyl-2*H*-benzopyran-8-yl)-3-(4-methoxyphenyl)-2-propen-1-one, and abyssinone-VI-4-O-methyl ether) having tremendous potential to exhibit antidiabetic activity by selectively modulating the promising therapeutic target PTP-1B that will prevent the degradation of insulin. Although, various electron withdrawing/donating groups such as –OH, –CH_3, –OCH_3, etc. in the pharmacophore have been perceived to play critical role in the interaction with the amino acid residues (particularly, Phe 182, Arg 47 and Asp 48) situated with the loop of tyrosine phosphate; however, no concrete prediction(s) has been laid toward the structure–activity relationships (SARs) for the 1,3-diphenyl-2*E*-propene-1-one-based PTP-1B inhibitors in this chapter which could have immense importance for the medicinal chemists due to several limitations. In modern days, these natural product chalcone-based PTP-1B inhibitors are not under clinical use and they have not received any such attention in modern-day medicine as they are not explored clinically in terms of toxicological profiles to develop a suitable formulation. In the near future, it is expected that these chalcone-based PTP-1B inhibitors will open new avenues of diabetotherapeutics.

KEYWORDS

- **chalcone**
- **diabetes mellitus**
- **antihyperglycemic**
- **hypoglycemic**
- **protein tyrosine phosphatase**
- **PTP-1B**
- **natural**
- **inhibitors**

REFERENCES

1. Shivhare, R. S.; Mahapatra, D. K. *Medicinal Chemistry-II*. ABD Publications Private Limited: Nagpur, 2019.
2. Godbole, M. D.; Mahapatra, D. K.; Khode, P. D. Fabrication and Characterization of Edible Jelly Formulation of Stevioside: A Nutraceutical or OTC Aid for the Diabetic Patients. Inventi Rapid: *Nutraceuticals* **2017,** *2017*(2), 1–9.
3. Kuhite, N. G.; Padole, C. D.; Amdare, M. D.; Jogdand, K. R.; Kathane, L. L.; Mahapatra, D. K. Hippuric Acid as the Template Material for the Synthesis of a Novel Anti-Diabetic 1,3,4-Thiadiazole Derivative. *Indian J. Pharm. Biol. Res.* **2017,** *5*(3), 42–45.
4. Gangane, P. S.; Kadam, M. M.; Mahapatra, D. K.; Mahajan, N. M.; Mahajan, U. N. Design and Formulating Gliclazide Solid Dispersion Immediate Release Layer and Metformin Sustained Release Layer in Bilayer Tablet for the Effective Postprandial Management of Diabetes Mellitus. *Int. J. Pharm. Sci. Res.* **2018,** *9*(9), 3743–3756.
5. Puranik, M. P.; Mahapatra, D. K. *Medicinal Chemistry-III*. ABD Publications Private Limited: Nagpur, 2019.
6. Mahapatra, D. K.; Chhajed, S. S.; Shivhare, R. S. Development of Murrayanine-Chalcone hybrids: An Effort to Combine Two Privilege Scaffolds for Enhancing Hypoglycemic Activity. *Int. J. Pharm. Chem. Anal.* **2017,** *4*(2), 30–34.
7. Mahapatra, D. K.; Dadure, K. M.; Haldar, A. G. M. Uracil Substitution on a Hippuric Acid Containing 1,3,4-Thiadiazole Scaffold: The Exploration of the Anti-Hyperglycemic Potential. *Int. J. Med. Sci. Clin. Res. Rev.* **2018,** *1*(2), 1–4.
8. Mahapatra, D. K.; Bharti, S. K. *Handbook of Research on Medicinal Chemistry: Innovations and Methodologies*. Apple Academic Press: New Jersey, 2017.
9. Mahapatra, D. K.; Bharti, S. K. *Medicinal Chemistry with Pharmaceutical Product Development*. Apple Academic Press: New Jersey, 2019.
10. Torrens, F.; Mahapatra, D. K.; Haghi, A. K., Eds. Biochemistry, Biophysics, and Molecular Chemistry: Applied Research and Interactions. Apple Academic Press: New Jersey, 2020.
11. Feldhammer, M.; Uetani, N.; Miranda-Saavedra, D.; Tremblay, ML. PTP1B: A Simple Enzyme for a Complex World. *Critic. Rev. Biochem. Mol. Biol.* **2013,** *48*(5), 430–445.
12. Bourdeau, A.; Dubé, N.; Tremblay, M. L. Cytoplasmic Protein Tyrosine Phosphatases, Regulation and Function: The Roles of PTP1B and TC-PTP. *Curr. Opin. Cell Biol.* **2005,** *17*(2), 203–209.
13. Tamrakar, A. K.; Maurya, C. K.; Rai, A. K. PTP1B Inhibitors for Type 2 Diabetes Treatment: A Patent Review (2011–2014). *Expert Opin. Ther. Pat.* **2014,** *24*(10), 1101–1115.
14. Tobin, J. F.; Tam, S. Recent Advances in the Development of Small Molecule Inhibitors of PTP1B for the Treatment of Insulin Resistance and Type 2 Diabetes. *Curr. Opin. Drug Discov. Dev.* **2002,** *5*(4), 500–512.
15. Tonks, N. k. PTP1B: From the Sidelines to the Front Lines!. *FEBS Lett.* **2003,** *546*(1), 140–148.

16. Zhang, Z. Y.; Lee, S. Y. PTP1B Inhibitors as Potential Therapeutics in the Treatment of Type 2 Diabetes and Obesity. *Exp. Opin. Invest. Drugs.* **2003,** *12*(2), 223–233.
17. Zhang, S.; Zhang, Z. Y. PTP1B as a Drug Target: Recent Developments in PTP1B Inhibitor Discovery. *Drug Discov. Today* **2007,** *12*(9–10), 373–381.
18. Jiang, C. S.; Liang, L. F.; Guo, Y. W. Natural Products Possessing Protein Tyrosine Phosphatase 1B (PTP1B) Inhibitory Activity Found in the Last Decades. *Acta Pharmacologica Sinica* **2012,** *33*(10), 1217.
19. Koren, S.; Fantus, I. G. Inhibition of the Protein Tyrosine Phosphatase PTP1B: Potential Therapy for Obesity, Insulin Resistance and Type-2 Diabetes Mellitus. *Best Pract. Res. Clin. Endocrinol. Metab.* **2007,** *21*(4), 621–640.
20. Taylor, S. D. Inhibitors of Protein Tyrosine Phosphatase 1B (PTP1B). *Curr. Top. Med. Chem.* **2003,** *3*(7), 759–782.
21. Goldstein, B. J. Protein-Tyrosine Phosphatase 1B (PTP1B): A Novel Therapeutic Target for Type 2 Diabetes Mellitus, Obesity and Related States of Insulin Resistance. *Curr. Drug Target Immune Endocr. Metab. Disorders.* **2001,** *1*(3), 265–275.
22. Wang, L. J.; Jiang, B.; Wu, N.; Wang, S. Y.; Shi, D. Y. Small Molecules as Potent Protein Tyrosine Phosphatase 1B (PTP1B) Inhibitors Documented in Patents from 2009 to 2013. *Mini Rev. Med. Chem.* **2015,** *15*(2), 104–122.
23. Huijsduijnen, R. H.; Wälchli, S.; Ibberson, M.; Harrenga, A. Protein Tyrosine Phosphatases as Drug Targets: PTP1B and Beyond. *Expert Opin. Ther. Targets* **2002,** *6*(6), 637–647.
24. Sun, J.; Qu, C.; Wang, Y.; Huang, H.; Zhang, M.; Li, H.; Zhang, Y.; Wang, Y.; Zou, W. PTP1B, a Potential Target of Type 2 Diabetes Mellitus. *Mol. Biol.* **2016,** *5*(4), 174.
25. Mahapatra, D. K.; Bharti, S. K. *Drug Design.* Tara Publications Private Limited: New Delhi, 2016.
26. Chhajed, S. S.; Upasani, C. D.; Wadher, S. J.; Mahapatra, D. K. *Medicinal Chemistry.* Career Publications Private Limited: Nashik, 2017.
27. Chhajed, S. S.; Bastikar, V.; Bastikar, A. V.; Mahapatra, D. K. *Computer Aided Drug Design.* Everest Publishing House: Pune, 2019.
28. Mahapatra, D. K.; Bharti, S. K.; Asati, V. Nature Inspired Green Fabrication Technology for Silver Nanoparticles. *Curr. Nanomed.* **2017,** *7*(1), 5–24.
29. Mahapatra, D. K.; Asati, V.; Bharti, S. K. Recent Therapeutic Progress of Chalcone Scaffold Bearing Compounds as Prospective Anti-Gout Candidates. *J. Crit. Rev.* **2019,** *6*(1), 1–5.
30. Mahapatra, D. K.; Bharti, S. K.; Asati, V. Anti-Cancer Chalcones: Structural and Molecular Target Perspectives. *Euro. J. Med. Chem.* **2015,** *98*, 69–114.
31. Mahapatra, D. K.; Bharti, S. K.; Asati, V. Chalcone Scaffolds as Anti-Infective Agents: Structural and Molecular Target Perspectives. *Euro. J. Med. Chem.* **2015,** *101*, 496–524.
32. Mahapatra, D. K.; Asati, V.; Bharti, S. K. Chalcones and their Therapeutic Targets for the Management of Diabetes: Structural and Pharmacological Perspectives. *Euro. J. Med. Chem.* **2015,** *92*, 839–865.
33. Mahapatra, D. K.; Bharti, S. K. Therapeutic Potential of Chalcones as Cardiovascular Agents. *Life Sci.* **2016,** *148*, 154–172.

34. Mahapatra, D. K.; Bharti, S. K.; Asati, V. Chalcone derivatives: Anti-inflammatory potential and molecular targets perspectives. *Curr. Top. Med. Chem.* **2017,** *17*(28), 3146–3169.
35. Mahapatra, D. K.; Bharti, S. K.; Asati, V.; Singh, S. K. Perspectives of Medicinally Privileged Chalcone Based Metal Coordination Compounds for Biomedical Applications. *Euro. J. Med. Chem.* **2019,** *174*, 142–158.
36. Mahapatra, D. K.; Asati, V.; Bharti, S. K. An Updated Patent Review of Therapeutic Applications of Chalcone Derivatives (2014-Present). *Exp. Opin. Ther. Pat.* **2019,** *29*(5), 385–406.
37. Mahapatra, D. K.; Asati, V.; Bharti, S. K. Natural and Synthetic Prop-2-ene-1-one Scaffold Bearing Compounds as Molecular Enzymatic Targets Inhibitors Against Various Filarial Species. In *Biochemistry, Biophysics, and Molecular Chemistry: Applied Research and Interactions*; Torrens, F., Mahapatra, D. K., Haghi, A. K., Eds.; Apple Academic Press: New Jersey, 2019.
38. Mahapatra, D. K.; Asati, V.; Bharti, S. K. Promising Anti-Cancer Potentials of Natural Chalcones as Inhibitors of Angiogenesis. In *Natural Products Chemistry: Biomedical and Pharmaceutical Phytochemistry*; Volova, T. G., Mahapatra, D. K., Khanna, S., Haghi, A. K., Eds.; Apple Academic Press: New Jersey, 2020.
39. Mahapatra, D. K.; Asati, V.; Bharti, S. K. Chalcone Scaffold Bearing Natural Anti-Gout Agents. In *Natural Pharmaceuticals and Green Microbial Technology: Health Promotion and Disease Prevention*; Mahapatra, D. K., Haghi, A. K., Eds.; Apple Academic Press: New Jersey, 2020.
40. Mahapatra, D. K.; Bharti, S. K.; Asati, V. Recent Perspectives of Chalcone Based Molecules as Protein Tyrosine Phosphatase 1B (PTP-1B) Inhibitors. In *Medicinal Chemistry with Pharmaceutical Product Development*; Mahapatra, D. K, Bharti, S. K., Eds.; Apple Academic Press: New Jersey, 2019.
41. Mahapatra, D. K.; Bharti, S. K., Asati, V. Recent Advancements in the Pharmacotherapeutic Perspectives of Some Chalcone Scaffold Containing Natural Compounds as Potential Anti-Virals. In *Natural Products Pharmacology and Phytochemicals for Health Care: Methods and Principles in Medicinal Chemistry*; Mahapatra, D. K., Aguilar, C. N, Haghi, A. K., Eds.; Apple Academic Press: New Jersey, 2020.
42. Mahapatra, D. K.; Bharti, S. K.; Asati, V. Chalcone (1,3-Diphenyl-2-Propene-1-One) Scaffold Bearing Natural Compounds as Nitric Oxide Inhibitors: Promising Anti-Edema Agents. In *Applied Pharmaceutical Practice and Nutraceuticals: Natural Product Development*; Mahapatra, D. K, Aguilar, C. N., Haghi, A. K., Eds.; Apple Academic Press: New Jersey, 2020.
43. Chen, R. M.; Hu, L. H.; An, T. Y.; Li, J.; Shen, Q. Natural PTP1B Inhibitors from Broussonetia papyrifera. *Bioorg. Med. Chem. Lett.* **2002,** *12*, 3387–3390.
44. Hoang, D. M.; Ngoc, T. M.; Da, N. T.; Ha, D. T.; Kim, Y. H.; Luon, H. V.; Ahn, J. S.; Bae, K. Protein Tyrosine Phosphatase 1B Inhibitors Isolated from Morus bombycis. *Bioorg. Med. Chem. Lett.* **2009,** 19, 6759–6761.
45. Zhang, L. B.; Lei, C.; Gao, L. X.; Li, J. Y.; Li, J.; Hou, A. J. Isoprenylated Flavonoids with PTP1B Inhibition from Macaranga Denticulata. *Nat. Prod. Bioprospect.* **2016,** 6, 25–30.

46. Na, M. K.; Jang, J.; Njamen, D.; Mbafor, J. T.; Fomum, Z. T.; Kim, B. Y.; Oh, W. K.; Ahn, J. S. Protein Tyrosine Phosphatase-1b Inhibitory Activity of Isoprenylated Flavonoids Isolated from Erythrina Mildbraedii. *J. Nat. Prod.* **2006,** *69*, 1572–1576.
47. Lei, C.; Zhang, L. B.; Yang, J.; Gao, L. X.; Li, J. Y.; Li, J.; Hou, A. J. Macdenti Chalcone, a Unique Polycyclic Dimeric Chalcone from Macaranga denticulata. *Tetrahedron Lett.* **2016,** *57*, 5475–5478.
48. Yoon, G.; Lee, W.; Ki, S.; Cheon, S. H. Inhibitory Effect of Chalcones and their Derivatives from Glycyrrhiza Inflate on Protein Tyrosine Phosphatase 1B. *Bioorg. Med. Chem. Lett.* **2009,** *19*, 5155–5157.
49. Park, S. J.; Choe, Y. G.; Kim, J. H.; Chang, K. T.; Lee, H. S.; Lee, D. S. Isoliquiritigenin Impairs Insulin Signaling and Adipocyte Differentiation through the Inhibition of Protein-Tyrosine Phosphatase 1B Oxidation in 3T3-L1 Preadipocytes. *Food Chem. Toxicol.* **2016,** *93*, 5–12.
50. Li, J. L.; Gao, L. X.; Meng, F. W.; Tang, C. L.; Zhang, R. J.; Li, J. Y.; Zhao, W. M. PTP1B Inhibitors from Stems of Angelica keiskei (Ashitaba). *Bioorg. Med. Chem. Lett.* **2015,** *25*, 2028–2032.
51. Ha, M. T.; Seong, S. H.; Nguyen, T. D.; Cho, W. K.; Ah, K. J.; Ma, J. Y.; Woo, M. H.; Choi, J. S.; Min, B. S. Chalcone Derivatives from the Root Bark of Morus alba L. Act as Inhibitors of PTP1B and α-Glucosidase. *Phytochemistry* **2018,** *155*, 114–125.
52. Ren, L.; Li, L. Z.; Huang, J.; Huang, L. Z.; Li, J. H.; Li, Y. M.; Tang, S. Y. New Compounds from the Seeds of Psoralea corylifolia with their Protein Tyrosine Phosphatase 1B Inhibitory Activity. *J. Asian Nat. Prod. Res.* **2019,** 1–6.

CHAPTER 8

Re-Highlighting the Potential Natural Resources for Treating or Managing the Ailments of Gastrointestinal Tract Origin

VAIBHAV SHENDE[1], SAMEER A. HEDAOO[1], MOJABIR HUSSEN ANSARI[1], POOJA BHOMLE[1], and DEBARSHI KAR MAHAPATRA[2*]

[1]*Department of Pharmaceutics, Gurunanak College of Pharmacy and Technical Institute, Nagpur 440026, India*

[2]*Department of Pharmaceutical Chemistry, Dadasaheb Balpande College of Pharmacy, Nagpur 440037, India*

[*]*Corresponding author. E-mail: mahapatradebarshi@gmail.com*

ABSTRACT

Gastrointestinal (GI) disorder is the term used to refer any condition or ailment that occurs within the gastrointestinal tract (also referred to as the GIT). Various natural drug (*Anethum graveolens* L., *Carum carvi*, *Cinnamomum tamala*, *Coriandrum sativum* L., *Foeniculum vulgare*, *Zingiber officinale* L., *Pinus roxburghii*, *Plumbago zeylanica*, *Punica granatum* L., *Saussurea lappa*, *Tamarindus indica*, and *Valeriana wallichii*) treatments are effective in lowering the signs and symptoms of GI disorders such as purposeful dyspepsia, constipation, and postoperative ileus, a painful situation that may additionally affect sufferers after a bowel surgery. Herbal medicines serve a valuable role in the management of patients with functional GI disorders. Herbal remedy can also help with the not unusual GI circumstance irritable bowel syndrome (IBS). Many of the medication used to deal with GI issues are useless or purpose side effect. Herbal remedy is a safe, holistic alternative that normally has no prominent adverse effect. From this chapter, a correspondence between ethnopharmacological knowledge with modern scientific findings (antioxidant, anti-inflammatory, anti-ulcer, gastroprotective, etc.) and data of validated experiments have been presented.

8.1 INTRODUCTION

Gastrointestinal (GI) disorder is the term used to refer any condition or ailment that occurs within the gastrointestinal tract (also referred to as the GIT). GIT is a series of hollow organs that form a long nonstop passage from mouth, esophagus, stomach, small intestine, large intestine, and ultimately the anus. The GIT, collectively with liver, pancreas, and gallbladder make up the complete digestive system.[1] A considerable network of blood vessels supply blood to these organs and also transports vitamins away to different organs in the body. Nerves and hormones work together to regulate the functioning of the digestive system and bacteria that live inside our GIT (referred to as our gut flora or microbiome) play a function in digestion, immunity, and our usual health. A membranous sac referred to as the peritoneum holds all the digestive system organs in place.[2]

8.2 COMMON GASTROINTESTINAL DISORDERS

GI diseases are found across a diverse range from new-born to elderly individuals. Both women and men suffer equally from these diseases. GI problems may be uncomfortable, disruptive, and embarrassing. A common disease is irritable bowel syndrome (IBS), which includes various symptoms, including abdominal pain, bloating, gas, diarrhea, and constipation. It has been estimated that 36% of patient registered in gastroenterology clinics suffer from IBS. Some diseases associated with the GI system are epidemiology surveillance diseases and they must be reported every year. These diseases consist of dysentery, cholera, food poisoning, mushroom poisoning, diarrhea, and hepatitis.[3] A number of different conditions or sicknesses can affect the GIT and feature an effect on digestion and/or our common health. Some situations have similar symptoms, and in addition medical investigation may be required earlier than a doctor arrives at a diagnosis. Common GI problems include:

8.2.1 CELIAC DISEASE

Celiac ailment is a serious autoimmune disorder in which the small gut is hypersensitive to gluten. Ingestion of gluten causes the immune device of the body to assault the small gut, main to damage to the villi of the small intestine, which are small fingerlike projections that sell nutrient absorption. These can begin at any age and symptoms encompass bloating, adjustments in bowel habit, rashes, weight reduction, and poor increase rate in children.[4]

8.2.2 CONSTIPATION

Constipation is the time period used to describe difficulty or infrequency in passing stools. Not everyone has each day bowel movement, so the passage of time between bowel motions earlier than constipation takes place varies from person to person. When anyone is constipated, their stools are commonly small, hard, dry, and hard to pass. Other symptoms may include bloating or distention in the stomach and pain during a bowel movement.[5]

8.2.3 CROHN'S DISEASE

It is a persistent bowel ailment that reasons patches of inflammation within the GIT anywhere among the mouth and the anus. Although the location in which the small intestine joins the massive intestine is most usually affected. The symptoms may include diarrhea that persist for numerous weeks, abdominal ache, and weight loss. Around 50% of people with Crohn's sickness note blood or mucus of their faces and some may file an argent need to move their bowels or sensation of incomplete evacuation.[6]

8.2.4 PEPTIC ULCER

Peptic ulcer disorder is an umbrella time period used to describe each gastric and duodenal ulcers, which might be mall holes that may occur inside the lining of your stomach (gastric ulcer) or upper part of your small intestine (duodenal ulcers). Duodenal ulcers are most common and are much more likely in people between 30 and 50 years. Gastric ulcers most customarily affect middle-aged or aged people. The most common motive is an infection with bacteria known as *Helicobacter pylori* that are commonly acquired in childhood, even though maximum people in no way expand ulcers. Symptoms typically include abdominal pain and heartburn.[7]

8.2.5 ULCERATIVE COLITIS

Ulcerative colitis affects most effectively the innermost lining of the colon. Although only a limited part of the bowel gets drastically affected, but, it is perceived that the whole of the colon remains inflamed. Symptoms like Crohn's disease encompassing diarrhea and the frequent need of bowel movement is often seen along with symptoms of rectal bleeding or bloody stools, stomach pain, tiredness, and lack of appetite. The reason remains unknown despite the fact that an extraordinary immune response seems answerable for the inflammation and weight-reduction plan and stress worsens the condition.[8]

8.2.6 GASTROESOPHAGEAL REFLUX DISEASE (GERD)

Gastroesophageal reflux disease (GERD) is also known as heartburn or acid reflux. It occurs whilst the hoop of muscle fibers that surrounds the doorway to our stomach (known as the decrease esophageal sphincter) turns weak and acts as a substitute of ultimate tightly closed to save the backflow of food back up. Esophagus, it remains partially open, allowing partly digested meals and belly acid to leak lower back up the esophagus, inflicting irritation. The primary signs associated with GERD are regurgitation, heartburn, chest pain, and nausea.[9]

8.3 MEDICINAL PLANTS FOR TREATING NUMEROUS GASTROINTESTINAL DISORDERS

If patients have tried conventional treatment without effect, natural alternatives may additionally help. Herbal medication can benefit humans experiencing GI issues that cannot be treated using traditional drug therapy, consistent with a study. Various natural drug treatments are effective in lowering the signs and symptoms of GI disorders such as purposeful dyspepsia, constipation, and postoperative ileus, a painful situation that may additionally affect sufferers after a bowel surgery. Herbal medicines serve a valuable role in the management of patients with functional GI disorders. Herbal remedy can also help with the not unusual GI circumstance IBS. Many of the medication used to deal with GI issues are useless or purpose side effect. In a few cases, this has led to capsules being withdrawn from the market. Herbal remedy is a safe, holistic alternative that normally has no prominent adverse effect.[10–11]

8.3.1 ANETHUM GRAVEOLENS L.

(Synonym: Pastinaca anethum spreng; family: Apiaceae)

It consists of dried leaves of the *A. graveolens* L. The fruit, leaf, and essential oil are the therapeutically active plant parts. It contains essential oils, fatty oil, moisture (8.39%), proteins (15.68%), carbohydrates (36%), fiber (14.80%), ash (9.8%), and mineral elements such as calcium, potassium, magnesium, phosphorous, sodium, vitamin A, and niacin. The fruits are acrid, bitter, thermogenic, deodorant, digestive, carminative, stomachic, anthelmintic, anodyne, anti-inflammatory, diuretic, emmenagogue, galactagogue, expectorant, cardiotonic, anaphrodisiac, febrifuge, sudorific, antispasmodic, anti-dysentery, alexiteric, and vulnerary. They are also useful in halitosis, flatulence, colic, dyspepsia, intestinal worms, odontalgia, arthralgia, inflammation, strangury, amenorrhea, fever, ulcer, and hepatopathy.

8.3.2 CARUM CARVI

(Synonym: Caraway Seed; family: Umbelliferae)

It consists of dried fruits of *Carum carvi*. The fruit is the therapeutically active plant part. The essential oil obtained from the whole plant is as follows: α-pinene, 0.2%; β-pinene, 0.3%; Camphene, 0.3%; myrcene, 1.5%; δ-3-carene, 1.0%; Limonene, 4.2%; γ-terpene, 2.7%; *p*-cymene, 0.3%; cadinene, 37.2%; myristicine, 1.2%; carvyl and dihydrocarvyl acetate, 1.1%; dihydrocarvone, 2.3%; carvone, 31.2%; terpinine-4-ol,

1,2-dihydrocarveol, 9.5%; etc. The extract of *C. carvi* is tested for its potential anti-ulcerogenic activity against indomethacin-induced gastric ulcers of the rat as well as for its antisecretory and cytoprotective activities. The extract produced a dose structured anti-ulcerogenic activity associated with a reduced acid output and a multiplied mucin secretion, a boom in prostaglandin E_2 release and a decrease in leukotrienes. The impact on pepsin content material is alternatively variable and did not appear to endure a relationship with the anti-ulcerogenic activity.

8.3.3 *CINNAMOMUM TAMALA*

(Synonym: Indian cassia lignea; family: Lauraceae)

It consists of dried leaves of *C. tamala*; a small evergreen tree found in the areas of Himalayas. The composition of volatile oil is as follows; α-thujene, 0.02%; α-pinene, 0.07%; β-pinene, 0.03%; 6-methyl-5-heptene-2-one, trace %; myrcene, 0.02%; 3-carene, 0.02%; *p*-cymene, 0.6%; camphor, trace %; etc. It is sweet, slightly penetrating, hot in potency, easy for digestion, cures diseases of kapha, vata, piles, nausea, and rhinitis. The bark is carminative.

8.3.4 *CORIANDRUM SATIVUM L.*

(Synonym: Coriander; family: Umbelliferae)

It consists of dried fruits of *C. sativum*; an annual, slender, glabrous herb. The fruit contains essential oil—linalool, linalyl, thymol, β-caryphyllene, α-pinene, borneol, limonene, β-phellendrene, citranellol. It also provides calorific contents such as protein 12.37 g, total lipid (fat) 17.77 g, carbohydrate 54.99 g, fiber 41.9 g, calcium 709 mg, sodium 35 mg, zinc 4.70 mg, trace amounts of vitamin A, vitamin C, vitamin D_2, and vitamin D_3, respectively. The plant showed the presence of essential oil, tannins, terpenoids, reducing sugars, alkaloids, phenolics, flavonoids, fatty acids, sterols, glycosides, etc. *C. satvum* is astringent, unctuous, diuretic, easily digestible, bitter, pungent, hot in potency, digestive, anthelmintic, improves taste, sweet after digestion, mitigates all the three doshas, thirst, burning sensation, vomiting, dyspnea, cough emaciation; fruit is aromatic, stimulant, carminative, stomachic, ant bilious, refrigerant, tonic, diuretic, and aphrodisiac. Fresh leaves are pungent and stomachic.

8.3.5 FOENICULUM VULGARE MILL

(Synonym: Fennel; family: Umbelliferae)

It consists of dried fruits of *F. vulgare*, which is an aromatic herb. The fruit contains volatile oil such as trans-anethole, fenchone, ocimene, β-monene, methyl chavicol, and α-phellandrene, sugar, moisture, protein, calcium, magnesium, vitamin A, vitamin E, vitamin K, vitamin C, folate, riboflavin, sodium, niacin, and thiamine. *F. vulgare* is also similar to properties of *A. sowa*, essentially it cures pain of the vagina, dyspepsia, good for heart, intestinal worms and destroys semen, causes dryness, hot in potency, digestive, cures cough, vomiting, mitigates kapha and vata. The volatile oil is bactericidal and antifungal, and has been shown to be effective *in vitro* against *Staphylococcus aureus* and *Candida albicans*.

8.3.6 ZINGIBER OFFICINALE L.

(Synonym: Ginger; family: Zingiberaceae)

It consists of dried rhizomes (raw as well as dry) of *Zingiber officinale* L. Essential oils, phenolic compound, flavonoids, carbohydrates, proteins, alkaloids, glycosides, saponins, steroids, terpenoids, tannin, monoterpene hydrocarbons, and oxygenated monoterpenes are present in the rhizomes. The raw ginger acrid, thermogenic, carminative, laxative, and digestive. It is useful in anorexia, vitiated conditions of vata and kapha, dyspepsia, pharyngopathy, and inflammations. The dry ginger is acrid, emollient, expectorant, anthelmintic. It is useful in dropsy, otalgia, cephalalgia, asthma, cough, colic, diarrhea, flatulence, etc.

8.3.7 PINUS ROXBURGHII

(Synonym: Chir Pine; family: Pinaceae)

It consists of dried leaves of the *Pinus roxburghii*. Wood, oleoresin, oil are the medicinally privileged parts. The extracts are known for the presence of alkaloids, glycosides, flavonoids, terpenoids, steroids, carbohydrates, proteins, fats, fixed oil, phenolic compounds, etc. The wood is acrid, bitter, sweet, emollient, aromatic, antiseptic, deodorant, hemostatic, stimulant, anthelmintic, digestive, liver tonic, diaphoretic, rubefacient, and diuretic. It is useful in hepatopathy, bronchitis, inflammation, skin disease, expectorant, halitosis, etc.

8.3.8 PLUMBAGO ZEYLANICA L.

(Synonym: Rosy-flowered leadwort; family: Plumbaginaceae)

It consists of dried roots of *Plumbago zeylanica.* It contains plumbagin, glucopyranoside, and sitosterol. Plumbagic acid glucosides; 3-*O*-beta glucopyranosyl plumbagic acid methyl ester along with five naphthoquinones are found to be present in roots. *P. zeylanica* is pungent both in taste and after digestion; is digestive, causes dryness, hot in potency, cures diseases of the duodenum, leprosy and other skin diseases, mitigates vata and kapha, and is water absorbent. It increases the digestive power and promotes appetite.

8.3.9 PUNICA GRANATUM

(Synonym: Pomegranate; family: Puniacacaceae)

It consists of dried seeds and flowers of *Punica granatum.* The plant contains apigenin, betulinic acid, callistephin, cyanthamine, conine, cyanidin and its diglucoside, ellagic acid, pelargonin, estradiol, pelletierine, polyphenols, and lipids. *P. granatum* is sweet, mitigates all three doshas, cures thirst, burning sensation, and fever, removes bad smell, increases semen, easily digestible, astringent, causes constipation, unctuous, bestows intelligence and strength.

8.3.10 SAUSSUREA LAPPA C.B. CLARKE

(Synonym: Costus root, Kuth; family: Asteraceae)

It consists of dried roots of *S. lappa.* Roots are used for medicinal purposes. The extracts obtained from the roots contain chemical constituents like dehydrocostus lactone, cynaropicrin, lappadilactone, germacrenes, betulinic acid, lignin glycoside, dihydro costunolide, alpha amorphenic acid, etc. *S. lappa* is hot in potency, pungent, sweet in taste, increase semen, bitter, easily digestible, cures herpes, cough, kushta, and mitigates kapha; essential oil possesses carminative, strong antiseptic, disinfectant properties and is an expectorant, diuretic and cardiac stimulant. The plant extract is known to raise blood pressure, relaxed muscles of intestine and uterus, antagonized spasmogenic action of acetylcholine and histamine in rats. It showed marked bronchodilator action in guinea pigs but maximum response was interior to that of epinephrine.

8.3.11 *TAMARINDUS INDICA*

(Synonym: Tamarind tree; family: Leguminosae)

It consists of dried fruits of *T. indica*. Roots, leaves, fruits, and seeds are the medicinally active parts. The fruits contain organic acid (10%) and their salts namely tartaric acid, citric acid, maleic acid, sodium and potassium tartrate (8%), and invert sugars (30–40%). The bark of *T. indica* is astringent, emmenagogue, tonic, anti-diarrheal, thermogenic, anthelmintic, anti-inflammatory, antifungal, diuretic, useful in vitiated conditions of vata, swelling, fever, gastropathy, wounds, ulcers, jaundice, scabies, and tumors. The seeds are digestive, carminative, cooling, aphrodisiac, stomachic, used in constipation and as tonic.

8.3.12 *VALERIANA WALLICHII*

(Synonym: Valerian; family: Valerianaceae)

It consists of dried roots of *V. wallichii*. The roots contain valepotriates (0.5–3%), essential oils (0.5–1.5%), valeriosidat, a water soluble iridoid ester glucoside, and alkaloids. *V. wallichii* is bitter, acrid, astringent, thermogenic, emollient, carminative, digestive, stomachic, hepatoprotective, cardiotonic, sedative, having depressant effect on CNS, and it is a good remedy for hysteria. It is useful in vitiated conditions of kapha and vata, ulcers, epilepsy, convulsions, and dyspepsia (Table 8.1).

TABLE 8.1 A Number of Well-Known Medicinal Plants Used in the Treatment of Gastrointestinal Disorders.

Scientific name (common name)	Picture	Family	Parts use	Traditional therapeutic effect	Other therapeutic effects in literature
Anethum graveolens L. (Anethum Arvense Salisb)		Apiaceae	Seeds, fruits	Digestive, carminative, stomachic, diuretic	Colic pain, flatulence, carminative, piles
Carum carvi (Meridian Fennel, Caraway)		Umbelliferae	Seeds, plant extracts, oils	Blood pressure lowering, anti-colitis, carminative, astringents	Diarrhea, dyspepsia, flatulent indigestion, Improves liver function
Cinnamomum tamala (Indian Cassia Lignea)		Lauraceae	Leaves	Gastroprotective, alleviate pain, inflammation, arthritic, rheumatism, antioxidant	Genitourinary, antioxidant

TABLE 8.1 *(Continued)*

Scientific name (common name)	Picture	Family	Parts use	Traditional therapeutic effect	Other therapeutic effects in literature
Coriandrum sativum L. (Coriander)		Umbelliferae	Leaves, fruits	Indigestion, carminative, inflammation, abdominal discomforts	Cough, bronchitis, diarrhea, folk medicine
Foeniculum vulgare Mill (Fennel)		Umbelliferae	Fruits	Digestive, anthelmintic, antioxidant, flatulence, inflammation	Antiviral, antibacterial, antifungal
Zingiber officinale L. (Ginger)		Zingiberaceae	Rhizomes	Indigestion, inflammation, loss of appetite, vomiting	Constipation, pain, cough, palpitation

TABLE 8.1 *(Continued)*

Scientific name (common name)	Picture	Family	Parts use	Traditional therapeutic effect	Other therapeutic effects in literature
Pinus roxburghii Sarg (Chir Pine)		Pinaceae	Wood, oleoresin, oil	Digestive, stimulant, intestinal, stomachic, inflammation	Piles, bronchitis, asthma, disease of the liver and spleen
Plumbago zeylanica (Chitrak)		Plumbaginaceae	Roots, leaves	Stomachic, antioxidant, carminative, cure intestinal trouble, digestive problems	Piles, helminthic, dysentery, bronchitis, itching
Punica granatum L. (Pomegranate)		Punicaceae	Roots, bark, seeds, flowers, fruits	Intestinal parasites, antioxidant, diarrhea	Hemorrhoids, treatment and prevention of cancer, diabetes

TABLE 8.1 *(Continued)*

Scientific name (common name)	Picture	Family	Parts use	Traditional therapeutic effect	Other therapeutic effects in literature
Saussurea lappa C.B. Clarke (Kuth, Costus)		Asteraceae	Roots, seeds, stem, flower, leaves	Digestive, colic, abdominal pain, hepatoprotective, antioxidant	Antiviral, antifungal, diabetics
Tamarindus indica (Tamarind)		Leguminosae	Fruits, seeds, roots, leaf	Anthelmintic, fruits digestive, gastropathy, antioxidant, hepatoprotective,	Diarrhea, laxative, inflammation
Valeriana wallichii (Indian Valerian)	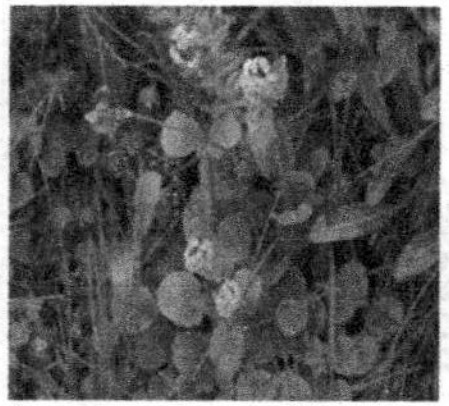	Valerianaceae	Plant extracts	Gastrospasm, hepatoprotective	Diarrhea, diuretic, hypertension

8.4 CONCLUSION

During the recent years, research has been focused on plants and their inherent properties of therapeutic utilities. These commonly available plants; namely, *Anethum graveolens* L., *Carum carvi*, *Cinnamomum tamala*, *Coriandrum sativum* L., *Foeniculum vulgare*, *Zingiber officinale* L., *Pinus roxburghii*, *Plumbago zeylanica*, *Punica granatum* L., *Saussurea lappa*, *Tamarindus indica*, and *Valeriana wallichii* have been examined to evaluate their efficacy as drugs to prevent and deal with a number of diseases associated with GIT. Phytotherapy has usually been beneficial for the person affected with GI diseases. From this chapter, a correspondence between ethnopharmacological knowledge with modern scientific findings (antioxidant, anti-inflammatory, anti-ulcer, gastroprotective, etc.) and data of validated experiments have been presented. It can also be concluded that investigation of new gastroprotective plants and identification of the herbal compound that they contain are essential consideration for the discovery of new drugs with less side effects, much less toxicity, less cost more efficacy in prevention and administration of exceptional gastric disorders.

KEYWORDS

- **gastrointestinal**
- **diseases**
- **disorders**
- **medicinal**
- **natural**
- **therapy**
- **herbal**

REFERENCES

1. Neamsuvan, O.; Ruangrit T. A Survey of Herbal Weeds that are Used to Treat Gastrointestinal Disorders from Southern Thailand: Krabi and Songkhla Provinces. *J. Ethnopharmacol.* **2017,** *196*, 84–93.
2. Gastrointestinal System Anatomy. https://healthengine.com.au.

3. https://www.slideshare.net/mobile/PoojaGoswami5/digestive-system-and-its-disease.
4. Parzanese, I.; Qehajaj, D.; Patrinicola, F.; Aralica, M.; Chiriva-Internati, M.; Stifter, S.; Elli, L.; Grizzi, F. Celiac Disease: From Pathophysiology to Treatment. *World J. Gastrointest. Pathophysiol.* **2017,** *8*(2), 27.
5. Leung, L.; Riutta, T.; Kotecha, J.; Rosser, W. Chronic Constipation: An Evidence-Based Review. *J. Am. Board Fam. Med.* **2011,** *24*(4), 436–451.
6. Baumgart, D. C. The Diagnosis and Treatment of Crohn's Disease and Ulcerative Colitis. *Deutsches ärzteblatt International*. **2009,** *106*(8), 123.
7. Kuna, L.; Jakab, J.; Smolic, R.; Raguz-Lucic, N.; Vcev, A.; Smolic, M. Peptic Ulcer Disease: A Brief Review of Conventional Therapy and Herbal Treatment Options. *J. Clin. Med.* **2019,** 8(2):179.
8. Collins, P.; Rhodes, J. Ulcerative Colitis: Diagnosis and Management. *BMJ* **2006,** *333*(7563), 340–343.
9. Clarrett, D. M.; Hachem, C. Gastroesophageal Reflux Disease (GERD). *Missouri Med.* **2018,** *115* (3), 214.
10. Natural Remedies for Gastrointestinal Problems: Herbal Medicine. https://www.amcollege.edu/blog/natural-remedies-for-gastrointestinal-problems-herbal-medicine.
11. Agrawal S. S.; Tamrakar, B. P.; Paridhavi, M. *Clinically Useful Herbal Drugs*. Ahuja Publishing House: Delhi, 2005.
12. Snafi Al Esmail, A. The Pharmacological Importance of Anethum Graveolens. A Review. *Int. J. Pharm. Sci.* **2014,** *6*(4), 11–13.
13. https://www.candidegardening.com/GB/plants/eaabf98455d9bdfef9c454fdb2bec271.
14. Khan Mohiyuddin, R.; Ahmad, W.; Ahmad, M.; Hasan, A. Phytochemical and Pharmacological Properties of Carum Carvi. *Eur. J. Pharm. Med. Res.* **2016,** *3*(6), 231–236.
15. https://www.lepetitherboriste.net/plantesus/caraway.html.
16. Kumar, S.; Vasudeva, N.; Sharma, S. Pharmacological and Pharmacognostical Aspect of Cinnamomum tamala Nees & Eberm. *J. Pharm. Res.* **2012,** *5*(1), 480–484.
17. https://www.healthcaretech.in/index.php?id_product=34&id_product_attribute=0&rewrite=cinnamomum-tamala-oil-tezpat&controller=product.
18. Snafi Al Esmail, A. A Review on Chemical Constituents and Pharmacological Activities of Coriandrum sativum. *IOSR J. Pharm.* **2016,** *6*(7), 17–42.
19. https://www.theseedcollection.com.au/Coriander.
20. Badgujar, B. S.; Patel, V. V.; Bandivdekar, H. A. Foeniculum vulgare Mill: A Review of its Botany, Phytochemistry, Pharmacology, Contemporary Application, and Toxicology. *Bio. Med. Res. Int.* **2014,** *3*(8), 1–31.
21. https://pfaf.org/user/plant.aspx?LatinName=Foniculum+vulgare.
22. Bodagh, N. M.; Maleki, I.; Hekmatdoost, A. Ginger in Gastrointestinal Disorders: A Systematic Review of Clinical Trials. *Wiley Food Sci. Nutr.* **2018,** *4*(7), 1–13.
23. https://www.medicalnewstoday.com/articles/265990.php#benefits.
24. Chaudhary, K. A.; Ahmad, S.; Mazumder, A. Protective Effects of Cedrus deodara and Pinus roxburghii on Experimentally Induced Gastric Ulcers in Rat. *Int. J. Pharm. Sci.* **2014,** *6*(4), 587–591.
25. https://www.naturalmedicinefacts.info/plant/pinus-roxburghii.html.

26. Vishnukanta, Rana C. A. Plumbago zeylanica: A Phytopharmacological Review. *Int. J. Pharm. Sci. Res.* **2011,** *2*(2), 247–255.
27. https://www.athreyaherbs.com/products/organic-chitraka-plumbagao-zeylanica.
28. Haque, N.; Sofi, G.; Ali, W.; Rashid, M.; Itrat, M. A Comprehensive Review of Phytochemical and Pharmacological Profile of Anar (Punica granatum Linn) A heavens Fruit. *J. Ayu. Herb. Med.* **2015,** *1*(1), 22–26.
29. https://nurserylive.com/buy-shrubs-plants-online-in-indica/cfp-punica-granatum-arakta-plants-in-india.
30. Elango, K. M.; Ponnusankar, S. Saussurea lappa (Kuth Root): Review of its Traditional Uses, Phytochemistry and Pharmacology. *Orient Pharm. Exp. Med.* **2012,** *12*, 1–9.
31. https://www.innovedaherbs.com/products/saussurea-lappa.
32. Zohrameena, S.; Mujahid, M.; Bagga, P.; Khalid, M.; Noorul, H.; Nesar, A.; Saba, P. Medicinal Uses & Pharmacological Activity of Tamarindus indica. *World J. Pharm. Sci.* **2017,** *5*(2), 121–133.
33. https://www.seedsdelmundo.com/product/tamarindus-indica/
34. Sundaresan, N.; Ilango, K. Review on Valeriana Species—Valeriana wallichii and Valeriana jatamansi. *J. Pharm. Sci. Res.* **2018,** *10*(11), 2697–2701.
35. https://www.google.com/amp/s/www.ayurtimes.com/tagara-valeriana-wallichii/amp/s

CHAPTER 9

Recent Advancements of Curcumin Analogs and Curcumin Formulations in Context to Modern Pharmacotherapeutics Perspectives

ANIMESHCHANDRA G. M. HALDAR[1], KANHAIYA M. DADURE[2], and DEBARSHI KAR MAHAPATRA[3*]

[1]*Department of Applied Chemistry, Priyadarshini Bhagwati College of Engineering, Nagpur 440024, India*

[2]*Department of Chemistry, J. B. College of Science, Wardha 442001, India*

[3]*Department of Pharmaceutical Chemistry, Dadasaheb Balpande College of Pharmacy, Nagpur 440034, India*

[*]*Corresponding author. E-mail: mahapatradebarshi@gmail.com*

ABSTRACT

For thousands of years in traditional medicines, excellent sources of pharmaceutical active ingredients are medicinal plants for the development of new drugs. Turmeric having the scientific name *Curcuma longa* belongs to the Zingiberaceae family which grows in the tropical and subtropical regions. A number of phytochemicals including curcumin, demethoxycurcumin, and bisdemethoxycurcumin are present in the roots of turmeric. The polyphenolic crystalline yellowish–orange colored curcumin is the active ingredient in the herbal remedy. In China and India, the use of turmeric in traditional medicines is very common till today. The use of curcumin from turmeric as a folk remedy continues today. This chapter comprehensively focuses on various reported pharmacologically active derivatives of curcumin and curcumin-based formulations.

9.1 INTRODUCTION

For thousands of years in traditional medicines, medicinal plants are excellent sources of pharmaceutical active ingredients for the development of new drugs.[1,2] Turmeric having the scientific name *Curcuma longa* belongs to the Zingiberaceae family which grows in the tropical and subtropical regions.[3] Distinctive flavor, yellow color, and poorly water-soluble powder was obtained from the roots of the plant.

The polyphenolic crystalline yellowish–orange colored curcumin is the active ingredient in the herbal remedy.[4] In China and India, the use of turmeric in traditional medicines is very common till today. Apart from medicine, "Indian saffron" turmeric has been used in cosmetics and dyeing fabric.[5] In the 14th century, European explorers introduced this turmeric spice to the Western world.[6] The use of curcumin from turmeric as a folk remedy continues today. In the ancient Indian medical system, Ayurveda, a piece of cloth covered with turmeric paste is used to treat infected eyes, to dress wounds, treat bites, burns, and various diseases related to skin.[7] This ancient remedy is also used to treat dental diseases, digestive disorders such as dyspepsia and acidity, indigestion, flatulence, ulcers, as well as to alleviate the hallucinatory effects of hashish and other psychotropic drugs.[8]

9.2 SOURCE OF CURCUMINOIDS

A number of phytochemicals including curcumin, demethoxycurcumin, and bisdemethoxycurcumin are present in the roots of *Curcuma longa* (Fig. 9.1). Curcumin is the most important therapeutic agent for various diseases. The fresh roots contain 2–5% of curcumin. The solubility of curcumin is highest in ethanol, DMSO, and acetone because of its hydrophobic nature. It shows absorption maxima around 420 nm. The appearance of yellow to deep red color is achieved by curcumin in acidic pH conditions.

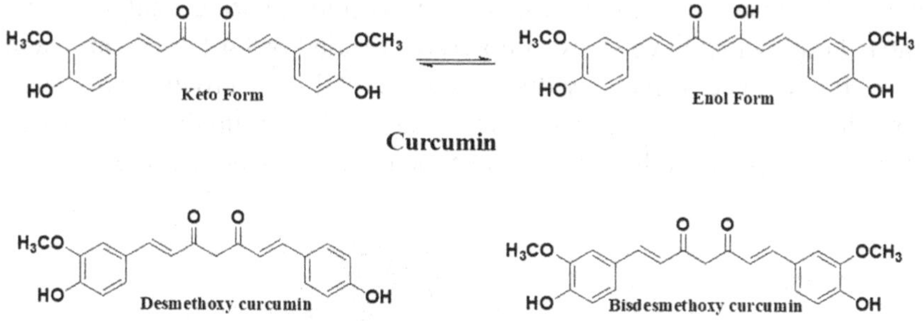

Keto Form

Enol Form

Curcumin

Desmethoxy curcumin

Bisdesmethoxy curcumin

FIGURE 9.1 Phytochemicals present in the roots of *Curcuma longa.*

9.3 PHARMACOLOGICAL ACTIVITIES

The accreditation of the therapeutic nature of turmeric is mainly due to curcuminoids that consist of curcumin and two related compounds, bis-demethoxycurcumin and demethoxycurcumin. Curcuminoids are commonly used as a dyeing agent as well as food additives apart from its utility as traditional medicine. The acceptable limit of curcuminoid intake as a food additive lies in the range of 0–3 mg/kg daily.

In India, oral hygiene[9] has been maintained by curcuminoids containing turmeric. It has traditionally been used for medical purposes for many centuries in countries such as India and China for the treatment of jaundice and other liver ailments.[10,11] The pharmacological activities of turmeric has explored in terms of curcuminoids such as antioxidant,[12] antiprotozoal,[13] antivenom activities,[14] antimicrobial,[15] antimalarial,[16] anti-inflammatory,[17] antiproliferative,[18] antiangiogenic,[19] antitumor,[20] and antiaging[21] properties (Fig. 9.2). It has also been used to treat ulcers, parasitic infections, various skin diseases, and anti-immune diseases, and to cure the symptoms of colds and flus.[22]

9.3.1 ANTICANCER ACTIVITY

Asymmetric curcuminoid analogs have been developed by Qiu et al.[23] and tested the NF-κB inhibitory potentials against gastric cancer cell lines. Effective inhibition of the growth of human gastric cancer cells like BGC-823, SGC-7901, and MFC cells through cell growth inhibition assays has been shown by some of the analogs. Different types of analyses like MTT assay, clonogenic assay, and Western blotting analysis were carried out, which have shown impressive results for further research and exerted anticancer effects by downregulating NF-κB activity. Irinotecan-combined analogs have shown their effects on NF-κB, inhibition of which effectively enhanced the sensitivity of the gastric cancer cell lines.

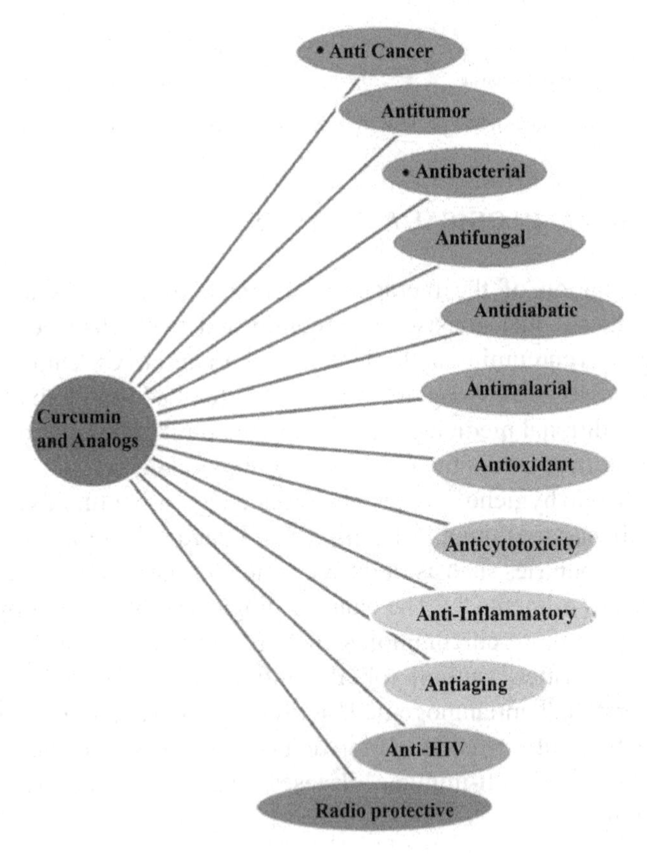

FIGURE 9.2 Biological activities of curcumin and various curcumin analogs.

Antiproliferative activity against lung adenocarcinoma cells H1975 and A549 has resulted from novel curcuminoid analogs established by structure–activity relationship (SAR), synthesized and studied by Gyuris et al.[24] by applying holographic microscopic imaging and viability assays. The TNF-α-induced NF-kB inhibition and autophagy induction effects correlated strongly with the cytotoxic potential of the analogs. In xenograft model, the most potent synthesized analogs combined with liposomes showed significant inhibition of the tumor growth in A549.

The comparative study of the influence of bis-demethoxycurcumin, demethoxycurcumin, and curcumin has been done by Yodkeeree et al.[25] on the expression of urokinase plasminogen activator metalloproteinases. The bis-demethoxycurcumin has more potential for an invasion in carcinoma cells than others by zymography analysis. All three curcuminoids have exhibited a significant effect on active metalloproteinases and urokinase plasminogen activator. The potency of antimetastasis measured by the downregulation of ECM degradation enzymes concluded that bis-demethoxycurcumin and demethoxycurcumin have invasiveness than curcumin.

Curcuminoid (*Z*)-3-hydroxy-1-(2-hydroxyphenyl)-3-phenylprop-2-en-1-one was synthesized by Ali et al.,[26] and MTT assay has provided the cytotoxicity test on breast cancer cell normal MCF-10A cell line and MCF-7 cell line. Meanwhile, flow cytometry has provided information about cell cycle regulation by cell population and induced apoptosis on MCF-7 cell. The regulation of cell cycle and apoptosis related to gene expression was analyzed by caspases activity analysis and quantitative real-time polymerase chain reaction. Measuring of reactive oxygen species (ROS) and glutathione (GSH) levels has concluded activation of oxidative stress on MCF-7.

Synthesized compound was found to possess selective cytotoxicity in breast cancer MCF-7 cell than normal MCF-10A cell. Annexin V/prodium iodide apoptosis analysis and cell cycle analysis by quantitation of DNA by flow cytometry reported that MCF-7 has been effectively seized by synthesized compounds at mitotic G2 phase, and after 72 h of incubation, induced apoptosis has observed. Upregulation of p53, transcription of p21, and downregulation of PLK-1 subsequently facilitate CDC2 phosphorylation which comes up with the arrest of the mitotic G2 phase.

Natural polyhydroxy polyphenolic compound with widely expressed approval of its antioxidant, anti-inflammatory, antipathogenic, and anticancer properties has been synthesized by Gurung et al.[27] However, low hydrophilicity, low cellular uptake, poor bioavailability, and rapid clearance are the main factors for its limited use. For evaluation of pharmacological properties of glycosides curcumin, the glycoside compound has been synthesized by chemoenzymatic glycosylation reactions in which UDP-α-D-2-deoxyglucose is the donor. The products have been obtained after glycosylation of curcumin as: curcumin 4′,4″-di-O-β-2-deoxyglucoside, curcumin 4′,4″-di-O-β-glucoside, curcumin 4′-O-β-2-deoxyglucoside, and curcumin 4′-O-β-glucoside. Synthesized products showed greater hydrophilicity and comparable antipathogenic activities. Additionally, the comparative study has provided information that the curcumin 4′-O-β-2-deoxyglucoside and curcumin 4′-O-β-glucoside showed more anticancer activities compared with the parent compounds.

The new compounds, bis-demethoxycurcumin and diacetylcurcumin, synthesized by Basile et al.[28] were more stable than curcumin which increased the activity in the internal medium and improved nuclear cellular uptake. The role in proliferation of the HCT116 cancer cells is very important in the chemotherapeutic effect mechanism. Both compounds showed the anticancer activity by damaging correct spindles formation, p21CIP1/WAF1-mitotic arrest, induced a p53 more favorably, and long effective for bis-demethoxycurcumin. The demethoxycurcumin was for anticancer effect on the human liver cancer cell line as tested by MTT assay flow cytometry and Western blotting assay.

A synthesized seven curcumin analogs have been reported by Ahsan et al.[29] for anticancer analysis in vitro. All the compounds were examined for their anticancer activity on nine different sections (colon cancer, renal cancer, leukemia, melanoma, non-small cell lung cancer, CNS cancer, ovarian cancer, prostate cancer, and breast cancer) represented by 60 NCI human cancer cell lines. The anticancer activity was shown by all the compounds in one dose assay concentration of 10 μM and, therefore, was evaluated further in five dose assays (0.01, 0.1, 1, 10, and 100 μM) and different dose parameters LC_{50}, TGI, and GI_{50} were calculated for each synthesized compound. The SR (leukemia) cell line with 0.03 μM (GI_{50}) showed the highest anticancer activity among different sections. Maximum activity on leukemia cell lines with GI_{50} values between 0.23 μM and 2.67 μM was shown by all the synthesized curcumin analogs.

Clinical significance of mammary serine protease inhibitor expression on IDCs breast cancer in the North Indian population was determined by Prasad et al.[30] and curcumin modulated its expression. Cell apoptosis occurred through the alteration of the expression of Bcl-2 and p53 proteins involved curcumin modulated mammary serine protease inhibitor expression in breast cancer cells. Curcumin mediated apoptosis mechanisms pointing involvement in the upregulation in the tumor growth can be suppressed by mammary serine protease inhibitor and metastasis in vivo and invasion in vitro in breast cancer and tumor cell motility.

New compounds series of curcuminoids having some replacement in C-5 position in curcumin with anticancer activity have been synthesized by Anthwal et al.[31] and screened against colon cancer (HCT116) cell lines and chronic myeloid leukemia (KBM5) human cancer cell line. Further, as compared with curcumin, these compounds also resulted in better inhibitory effect on TNF-5-induced NF-5B activation. Newly synthesized compounds were resulted to show better cytotoxicity than curcumin standard against cancer cell lines.

The combined Curcumin with carnosic acid has tested the effect of growth on human breast cancer cells by Einbond et al.[32] Significant effects have been measured by treating breast cancer cells with carnosic acid individuals or combined with curcumin on gene expression and cell proliferation. Carnosic acid showed proliferation of ER-negative breast cancer cells inhibition and selected for Her2 overexpressing cells and can thus have cancer stem cell inhibition. Combined carnosic acid and curcumin can play a major role in the prevention and treatment of triple-negative breast cancer.

Photodynamic therapy (PDT) as a therapeutic modality has developed by Lin et al.,[33] performance based on induced cell death via the formation of ROS under illumination. The comparative study showed higher antiproliferative action on breast cancer cells by demethoxycurcumin-PDT than curcuminoids-PDT significantly. In PDT cancer therapy, demethoxycurcumin worked as an effective photosensitizer. In inhibition of cell viability in breast cancer cell lines combined demethoxycurcumin-PDT treatment has been more sensitive than an individual alone.

Balaji et al.[34] have synthesized oxovanadium(IV) complexes, viz., [VO(Fc-tpy)(Curc)](ClO_4), [VO(Fc-tpy)(bDHC)](ClO_4), [VO(Fc-tpy)(bDMC)](ClO_4) and [VO(Ph-tpy)(Curc)](ClO_4), of 4′-ferrocenyl-2,2′:6′,2″-terpyridine (Fc-tpy), and 4′-phenyl-2,2′:6′,2″-terpyridine (Ph-tpy) and monoanionic curcumin (Curc), bis-dehydroxycurcmin (bDHC), and bis-demethoxycurcumin (bDMC) were prepared, characterized, and their photo-induced DNA cleavage activity through visible light was studied. The complexes as binders to calf thymus DNA showed the formation of -OH radicals by photocleavage of plasmid DNA in the red light of 647 nm. The complexes have been uptaken by the cell, estimated within 4 h of treatment, and measured by fluorescence microscopic. The photocytotoxicity has been shown by complexes in visible light range in HeLa cell lines and HepG2 cancer cells with low toxicity.

The effects of curcumin on HIF-1α in cisplatin has been examined by Ye et al.[35] The result showed that they were very sensitive to A549 and resistant to A549/DDP cell lines by RTPCR and Western blot on the basis of curcumin as a chemo-sensitizer in lung cancer. Combined cis-platinum-curcumin treatment clearly resulted in the inhibition of A549/cis-platinum cell proliferation, reversed cis-platinum resistance by promoting HIF-1α degradation. The anticancer strategies of curcumin through reducing HIF-1α–dependent P-gp might be a potent way for overcoming multidrug resistance. This combined effect of different chemotherapeutic agents with curcumin can be used as a promising favorable strategy for lung cancer treatment.

For improving the biological activity, 16 novel curcuminoid combined cinnamic acid derivatives have been synthesized by Feng et al.[36] Almost all the new compounds showed encouraging activity. Synthesized compounds showed much better antioxidant activity than vitamin C, and also had many-fold better antibacterial activity than ampicillin with a MIC of 0.5 μg/mL against Gram-positive cocci such as *S. aureus* and *S. viridans*. The compound showed the greatest anticancer activity with a much lower IC_{50}, which was 0.51 μM against MCF-7, 0.58 μM against HepG-2, 0.63 μM against LX-2, and 0.79 μM against 3T3 indicated that these compounds have potential for the treatment of cancer.

The newly curcumin derivatives containing glucosyl molecules have been synthesized by Rao et al.[37] for evaluating their biological activities. Increased tyrosinase enzyme inhibition and cytotoxic activity have been shown by glucosyl curcuminoids. The enhanced biological activity because of enhanced hydrophilicity of the glucosides containing curcuminoids was compared to the parent curcuminoids. Partition coefficient studies clarified the enhanced hydrophilicity of the synthesized compounds. Thus, the results showed a promising future of synthesized glucosyl curcumin as a therapeutic agent for anticancer.

Newly synthesized curcuminoids conjugated with methionine-substituted and selenomethionine-substituted (series 1) ferulic acid and caffeic acid conjugates (series 2), and glycyrrhetinic acid conjugates (series 3) with enhanced bioactivity have been synthesized by Cao et al.[38] The single compound of methionine-substituted and selenomethionine-substituted curcuminoids resulted in the most enhanced activity with a MIC value of 0.5 μM/mL against selected Gram-positive and Gram-negative. The derivatives exhibited remarkable results in an antioxidant test with an IC_{50} 2.4-fold to 9.3-fold lower when compared with curcuminoids. Antiproliferative activity of the synthesized derivatives showed higher effect against LX-2, Hep-G2, MDA-MB-231, and SMMC-7221 than parent compounds with an IC_{50} ranging from 0.18 μM to 4.25 μM.

A pyrazole conjugated curcuminoids derivatives series have been synthesized by Puneeth et al.[39] The effects of compounds were screened on different cancer cell lines by MTT assay for cancer cell inhibition. The analogs resulted in the growth inhibition effect on K562, HeLa, and MCF-7 cell lines with prominent IC_{50} values. Compound [4,4-(1*E*,1*E*)-2,2-(1-(3-chlorophenyl)-1*H*-pyrazole-3,5-diyl)bis(ethene-2,1-diyl)bis(2-methoxy phenol] exhibited a high degree of cytotoxicity against cancer cells and minimal growth inhibition effects against normal cells HEK293T and cell cycle analysis. Effective inhibition of cell cycle progression and decreased membrane potential in a minimum concentration-manner examined for the above compound through fluorescence shifting from red to green. Effective inhibition of cell arrest suggested that compound could be a promising anticancer agent.

Novel series of curcuminoid analogs bearing pyrazole/pyrimidine has synthesized by Ahsan et al.[40] According to the screening protocol, different 60 cell lines have been examined for anticancer effects. Some of the newly structured synthesized compounds showed improved anticancer activity in different assays. The most active compound of the series was 3,5-bis(4-hydroxy-3-methylstyryl)-1*H*-pyrazole-1-yl(phenyl)methanone which showed a mean growth percent of −28.71 in a single-dose assay.

9.3.2 ANTIOXIDANT ACTIVITY

Curcumin and 10 new derivatives (five derivatives of different substitutes, one isoxazole derivative, three pyrazole derivatives, and two other azole derivatives) have been amalgamated by Lozada-García et al.[41] and evaluated as antioxidant agents. The primary screening showed valuable results against different cancer cell lines (SKLU-1, HCT-15, K562, PC-3, and U-251MG), initiated us to measure the IC_{50} values for the potent compounds against K562 and HCT-15 cell lines. Thiobarbituric acid reactive substances assay antioxidant properties showed by the compound than parents, by α-tocopherol and quercetin as reference.

Zheng et al.[42] synthesized five series of curcuminoids and evaluated their antioxidant activity in vitro. The compounds with pyrazoline ring containing N1 thiocarbamoyl substituent resulted in the highest antioxidant properties among the synthesized compounds with inhibition percentage of 84.52 and 88.17%. This also resulted that electron-donating groups were necessary for enhancing the free radical scavenging properties; especially the phenolic hydroxyl group is a necessary component.

Fourteen curcumin analogs containing 3,4-dihydropyrimidine derivatives have been designed and synthesized by Sahu et al.[43] and their antioxidant activity has been evaluated. HCT-116, QG-56, and HeLa cell lines have been chosen for cytotoxicity screening by MTT assay method. The synthesized pyrimidines of curcumin exhibited enhanced antioxidant properties than base molecules.

Mošovskáa et al.[44] have performed the synthesis of curcumin cyclodextrin and curcumin–phospholipid complexes. Since neutral curcumin is characterized as poor hydrophilicity, the antioxidant activity of isolated curcuminoids was assessed by two methods (ABTS and FRAP assay) and their free radical scavenger activities of isolated were examined with those of synthesized complexes. The quercetin showed the highest ability to reduce ABTS radical cation whereas curcumin–phospholipid complex showed the lowest reducing ability. The quercetin has shown more reducing potential of tested samples than curcumin–phospholipid complex.

The antioxidant properties of the new seven synthesized compounds have been examined by Sokmen et al.[45] through two assays, β-carotene/linoleic acid and DPPH. It was found that, in general, the free radical scavenging ability of compounds was concentration-dependent. Higher free radical scavenging values were showed by compounds that contained *p*-OH phenolic groups than BHT which resulted that synthetic analogs are more potent antioxidants than BHT. The inhibition of the generation of conjugated dienes confirmed its free radical scavenging activity of hydroxy phenolic compound by using β-carotene-linoleic acid assay.

Sahu et al.[46] has structured and synthesized a distinct number of curcumin derivatives viz. benzylidene, pyrazole, and 4*H*-pyrimido [2,1-b] benzothiazole derivatives for their toxicity to the cell and free radical scavenging activity. The better antioxidant properties were showed by derivatives of curcumin than parent due to the potential of free radical capturing ability by the hydroxyl group. These curcumin derivatives can play a specific role as a promising antioxidant.

A number of asymmetric curcumin analogs and symmetric curcumin derivatives have been synthesized by Li et al.[47] The antioxidant activity of synthesized analogs was evaluated by DPPH assay, ABTS assay, TRAP assay, and NET assay, and antiproliferative activities of these analogs were assessed against the MCF-7, SMMC-7721, and PC-3 cell lines. The antioxidant activity showed by most of the asymmetric compounds was higher when they were compared with Vitamin C. Free radical scavenging activity was shown by all the synthesized compounds as compared with curcumin. Reduction of the O–H bond dissociation enthalpy increases the antioxidant activity that has been achieved by the shortening of the carbon chain.

Novel curcuminoids mimics have been synthesized with various heterocyclic, alicyclic amides of 3-aminoacetophenone, and aromatic by Bandgar et al.[48] and tested for different activities. Newly synthesized screened for antioxidant activity and xanthine oxidase inhibition have shown better results than the parents. Some compounds were resulted to be cancer cell lines inhibitors.

Li et al.[49] synthesized Ferrocenyl-substituted curcumin derivatives such as 1,7-bis(*p*-hydroxy-m-methoxyphenyl)-4-ferrocenylidene-hepta-1,6-diene (FCU), 1-(*p*-hydroxy-*m*-methoxyphenyl)-3-hydroxy-7-ferrocenylhepta-1,4,6-trien-5-one (FFT), and 1-(*p*-hydroxy-*m*-methoxyphenyl)-5-ferrocenyl-penta-1,4-diene-3-one (FDZ) and examined their free radical scavenger activities in trapping 2,2′-azinobis(3-ethylbenzothiazoline-6-sulfonate) cationic radical (ABTSþ), and hydroxy free radical-induced DPPH, Cu2þ/GSH. The experiment resulted that ferrocenyl group attached curcumin enhanced the antioxidant ability of the FCU, FFT, and FDZ.

Parvathy et al.[50] synthesized curcumin–amino acid conjugates by the reaction of t-Boc-protected amino acids with Curcumin where high free radical scavenger activity has been shown by the amino acid conjugate than the parent curcumin because of the substituted phenolic position of curcumin. Conjugates of curcumin with alkyl-substituted amino acids and the curcumin–cysteine conjugate have shown much superior antioxidant activity than curcumin.

9.3.3 ANTITUMOR ACTIVITY

Frederik Roos et al.[51] investigated that the naturally occurring curcumin has low bioavailability but light exposure has been shown to improve the bioavailability of curcumin. They investigated that light exposed curcumin increases efficacy in cell lines. TCCSUP, UMUC3, and RT112 cells were preincubated with low concentrations curcumin and then lighted to visible light 1.65 J/cm^2. Cell growth, cell proliferation, apoptosis, cell cycle progression were investigated. Cell proliferation, apoptosis, and tumor cell growth were strongly arrested when visible lighted curcumin was used. Light exposed Curcumin caused inhibition in different phases: RT112 in G_0/G_1 phase, TCCSUP in G_2/M phase, and UMUC3 in S-phase. Light exposure on curcumin increases the potential of antitumor on cancer cells.

Curcumin has an important property as a binding inhibitor of CBR1, was first reported by Hintz Peter et al.[52] Curcumin worked as a cofactor by blocking the CBR1 binding sites that affect daunorubicinol formation. Curcumin can leave the potential to enhance the efficiency of daunorubicin through the inhibition of CBR1 mediated reduction of daunorubicin to daunorubicinol. The inhibitory action of CBR1 can enhance the efficiency of daunorubicin in cancer tissue.

The compounds containing antitumor properties have been identified by Jiang et al.[53] from *C. longa* on HeLa cells were measured using an MTT (3,4,5-dimethyl thiazol-2-yl)-2,5-diphenyl tetrazolium bromide assay based on the composition and activity relationship. The antitumor activity of curcuminoids was significantly correlated through different correlation analysis.

9.3.4 ANTIDIABETIC ACTIVITY

Dahan et al.[54] have studied an effective preparation of nanoparticle colloidal dispersion of curcumin to improve bioavailability (THERACURMIN). THERACURMIN has high solubility in water and has better stability without any unpleasant odor. The product showed better bioavailability because natural curcumin has low bioavailability and the serum levels were 27.3 fold higher than that observed for 30 mg of curcumin. Curcumin has been shown to exhibit aldose reductase inhibitory activity but poor bioavailability coupled with fast metabolic elimination makes the compound therapeutically less effective.

Kondhare et al.[55] have developed novel aldose reductase inhibitors (ARIs) of therapeutic significance. The nanodispersion of curcumin

THERACURMIN tested for better ARI activity and evaluated the product for aldose reductase inhibition. Aldose Reductase, the key enzyme involved in the first and rate-limiting step of polyol pathway has been implicated in the development of late microvascular complications such as nerve-damaging due to diabetic, cataract, nephropathy, etc. As a consequence, the inhibition of this enzyme is of therapeutic significance to reduce the seriousness of chronic diabetic complications. It is evident from studies that THERACURMIN as expected exhibited excellent ARI activity in 20–30 μM range with IC_{50} of 3 μM. This is remarkable as the product contains only 15% curcuminoids in the total dispersion and curcumin itself displays potent aldose reductase inhibition activity.

Zeinab Ghorbani et al.[56] investigated the effect of curcumin which is a main constituent of turmeric. Curcumin lowered the glucose level in blood by reducing inflammatory state induced by hyperglycemia and hepatic glucose production, PPAR ligand-binding efficiency, activation of AMP kinase, and pancreatic cell function enhancement. Insulin sensitizer effects and antihyperglycemic were also characterized by curcumin molecules.

A different type of curcumin analogs with mono carbonyl moiety have been synthesized by Yuan et al.[57] and pharmacokinetic of curcumin is decided by beta-diketone moiety which has been removed from newly synthesized compounds. Some of the synthesized analogs inhibited rodent and human cortisone reductase. The synthesized analogs showed 4–20 times more inhibitory effects than parents. Curcumin analogs are highly selective, favorable for cortisone reductase, and have weak inhibition for 11β-HSD2. The previously tested compounds have potential therapeutic agents for type-2 diabetes by targeting 11β-HSD1.

Curcuminoids have a significant role in animal models of diabetes, both by lowering blood glucose levels and enhancing the long-term complications of diabetes has investigated by Teayoun et al.[58] They examined effective antidiabetic mechanisms of tetrahydro curcuminoids and curcumin. DPP4/α-glucosidase inhibitory activity resulted that curcumin analogs did not affect intestinal glucose metabolism, receptor tyrosine kinase activity, or 2-deoxyglucose directly. Curcumin analogs effectively suppressed Hep3B human hepatoma cells and dexamethasone-induced PEPCK. 5′-adenosine monophosphate-activated protein kinase phosphorylation enhanced by curcumin analogs and acetyl-CoA carboxylase as a downstream target in Hep3B and H4IIE cells with higher potency compared with metformin.

9.3.5 ANTIMALARIAL ACTIVITY

The antimalarial properties of two different series of curcumin analogs have been studied by Dohutia et al.[59] based on correlation antimalarial activity with their docking with *Pf*ATP6. The compounds were synthesized with retaining parent curcumin nucleus functional groups but new analogs changed in carbon chain length, unsaturated groups, and the increased ketonic groups. The compounds (1*E*,4*E*)-1,5-bis(4-methylphenyl)penta-1,4-dien-3-one,(1*E*,4*E*)-1,5-bis(4-methoxyphenyl)penta-1,4-dien-3-one and (*E*)-1,3-bis(4-hydroxylphenyl)prop-2-en-1-one showed IC_{50} values of 1.642 μM, 1.764 μM, and 2.59 μM in 3D7 strain and 3.039 μM, 7.40 μM, and 11.3 μM in RKL-2 strain, respectively. The antimalarial activity of synthesized compounds has been revealed through hydrophobic interaction with the residue Leu268 of the *Pf*ATP6 protein, through SAR studies.

The antimalarial activity of six curcumin derivatives has been initiated to assess by Dohutia et al.[60] based on binding affinities and correlation with docking and in vitro analysis of antimalaria. A library synthesis of 32 different curcuminoids was structured and *Pf*ATP6 protein has been used for docking. Among 32 compounds, 6 compounds having the best binding affinity were prepared and screened for their antimalarial evidence by using 3D7 strain of *Plasmodium falciparum*. ADME-Tox, pharmacodynamics, and pharmacokinetic profiles of the synthesized compounds were examined and reported. 4-FB was found to have similar binding energy to the standard artemisinin (–6.75 and –6.73 respectively) while 4-MB, 3-HB, 2-HB, B, 4-NB displayed better binding energy than curcumin (–5.95 kcal/mol, –5.89 kcal/mol, −5.68 kcal/mol, −5.35 kcal/mol, −5.29 kcal/mol, and −5.25 kcal/mol, respectively). The synthesized compounds with a concentration of 50 μg/mL showed best schizont inhibition while at low dose 5 μg/ml, only five synthesized compounds resulted in better inhibition than curcumin.

Newly synthesized 19 compounds have been evaluated by Balaji et al.[61] for their antimalarial screening in vitro. The antimalarial data concluded that all synthesized compounds with MKCs range between 3.87 and 25.35 μM showed parasiticidal activity and schizonticidal activity with IC_{50} range between 1.48 and 23.09 μM. One of them showed the most significant result with maximum schizonticidal (IC_{50}; 1.48 ± 0.10 μM) and parasiticidal activities (MKC; 3.87 ± 0.36 μM) could be identified as a promising lead for further investigations.

A number of novel monocarbonyl curcuminoids have been structured and synthesized by Manohar et al.[62] and tested for their activity against HeLa, KB, DU145, Molt4, and PC3 cancer cell lines. Six curcuminoid analogs showed potent cytotoxicity toward cell lines with IC_{50} values 1 μM, which is better than approved drug doxorubicin. In vitro analysis of antimalarial screening, synthesized analogs have been reported active against both CQ-resistant and CQ-sensitive of *P. falciparum*.

The effects of curcuminoids on the inhibition of *P. falciparum* in cell culture at fewer concentrations have been investigated by Shukla et al.[63] Previous studies have suggested that Ca^{2+}-ATPase (*Pf*ATP6) of *P. falciparum* is the target of many antimalarial drugs. However, the mechanism of inhibition of Ca^{2+}-ATPase (*Pf*ATP6) is not known. Through study, this is not confirmed that which isomer of curcuminoid is the most effective antimalarial compound. We address this issue using bioinformatics tools. Molecular model of Ca^{2+}-ATPase (PfATP6) of *P. falciparum* and docking of curcuminoid analogs of Zinc database of compounds (zinc.docking.org). The determination of the binding feasibility of 351 analogs has been done by docking software FlexX and Glide. The 20 synthesized curcuminoids have a good binding affinity with *Pf*ATP6 than curcumin. Both polar interactions and hydrophobic nature are the main factors for the binding affinity of curcuminoids to *Pf*ATP6. Their results suggested that synthesized compounds lead to the betterment of antimalarial drugs.

9.3.6 CYTOTOXICITY

The antifungal activity of *Curcuma longa* was demonstrated by Chen et al.[64] where alcohol extract of *Curcuma longa* was examined for a large number of fungi (*Fusarium chlamydosporum, Fusarium oxysporum, Rhizopus oryzae, Botrytis cinerea, Cladosporium cladosporioides, Fusarium tricinctum, Alternaria alternate, Sclerotinia sclerotiorum, Fusarium culmorum, Colletotrichum higginsianum*, and *Fusarium graminearum*) and EC_{50} values were reported. Inhibitory effects of the phytoconstituents; isocurcumenol, curzerene, germacrone, curcumenol, curcumin, curdione curcumol, and β-elemene derived were tested along with curdione complexes. During the comparison of differential proteomics of *F. graminearum*, some reproducible protein marks were spotted. Among these spots, some of the reproducible proteins are important in tRNA synthesis, glucose metabolism, and energy metabolism. Furthermore, several fungal physiological differences were also analyzed. The inhibition of fungal infection involves respiration, inhibition of ergosterol synthesis, and fungal cell membrane disruption.

A new series of triazole containing curcuminoids where the triazole ring is placed in the position of the dicarbonyl moiety was made as a possible strategy to enhance potency and selectivity by Caprioglio et al.[65] As compared to Curcumin, a proof-of-principle library of 28 compounds were tested for their cytotoxicity against SY-SY5Y and HeLa cells for their ability to inhibit NF-kB. Investigators found that some compounds have the ability to inhibit NF-kB without showing cytotoxicity while others activated NF-kB signaling.

Fifteen newly curcuminoids have been synthesized by Shi et al.[66] for inhibition against two human prostate cancer cell lines: PC3 and LNCaP. Twelve Different analogs are conjugated with flutamide like moiety in curcumin and methyl curcumin. Cell morphology study reveals that cytotoxicity of curcuminoids analogs reported with both prostate cancer cell lines PC3 and LNCaP effective because of the effect on pseudopodia formation.

A newly curcumin analog has been synthesized by Zhang et al.[67] having phenyl with different para-substituents. The different para-substituents have been examined for cytotoxicity against a large number of human cancer cell lines in vitro. Improved cytotoxicity against cell line A-431 and cell line U-251 have been reported by synthesized analogs showed effectiveness as a chemotherapeutic agent for skin cancer and glioma.

A new range of curcumin derivatives has been synthesized by Ferrari et al.[68] containing glycosylation of aromatic ring of curcuminoids to improve the fundamental characteristics of drug bioavailability through water solubility with greater kinetic ability. The glycosylation of curcuminoids decreases the cytotoxic effect of the synthesized compounds toward cisplatin (cDDP) which is resistant and sensitive to ovarian carcinoma cell lines. The glycoside conjugated curcuminoids reported much less toxicity to nontumorigenic cells.

9.3.7 ANTI-INFLAMMATORY ACTIVITY

A new range of curcumin analogs conjugated with asymmetrical pyrazole have been synthesized by Jadhav et al.[69] In vitro screening of all the curcuminoids have been performed for their anti-inflammation activity. Among the tested series, compounds 4-fluoro-*N*-(3-{3-[3-(4-methoxyphenyl)-1-phenyl-1*H*-pyrazol-4-yl]-acryloyl}-phenyl)-benzamide and *N*-(3-{3-[3-(4-chlorophenyl)-1-phenyl-1*H*-pyrazol-4-yl]-acryloyl}-phenyl)-4-fluoro-benzamide reported excellent hydrogen peroxide capturing activity as compared to standard BHT. Compounds *N*-(3-{3-[3-(4-chlorophenyl)-1-phenyl-1*H*-pyrazol-4-yl]-acryloyl}-phenyl)-4-fluoro-benzamide, 4-fluoro-*N*-{3-[3-(1-phenyl-3-*p*-tolyl-1*H*-pyrazol-4-yl)-acryloyl]-phenyl}- benzamide, 4-fluoro-*N*-(3-{3-[3-(3-nitro-phenyl)-1-phenyl-1*H*-pyrazol-4-yl]-acryloyl}-phenyl)-benzamide, and *N*-{3-[3-(1,3-diphenyl-1*H*-pyrazol-4-yl)-acryloyl]-phenyl}-4-fluoro-benzamide showed good DPPH free radical capturing activity. All the compounds showed better nitric oxide (NO) scavenging activity than standard ascorbic acid. In the in vitro testing, all the targeted compounds were examined for screening of their anti-inflammatory potency. Compounds *N*-(3-{3-[3-(4-chlorophenyl)-1-phenyl-1*H*-pyrazol-4-yl]-acryloyl}-phenyl)-4-fluoro-benzamide,4-fluoro-*N*-{3-[3-(1-phenyl-3-*p*-tolyl-1*H*-pyrazol-4-yl)-acryloyl]-phenyl}-benzamide, 4-fluoro-*N*-(3-{3-[3-(3-nitro-phenyl)-1-phenyl-1*H*-pyrazol-4-yl]-acryloyl}-phenyl)-benzamide, and *N*-{3-[3-(1,3-diphenyl-1*H*-pyrazol-4-yl)-acryloyl]-phenyl}-4-fluoro-benzamideshowedbetteranti-inflammation than the standard drug diclofenac sodium.

Three newly dimethylamino curcuminoids have been prepared by Banuppriya et al.[70] The arylidene curcuminoids and pyrazole curcuminoid derivatives were examined for their anti-inflammatory in vitro and antibacterial activities. Excellent anti-inflammatory properties have been shown by dimethylamino curcuminoids derivatives than their parent compounds. The newly synthesized curcuminoids derivatives showed an excellent inhibition of cyclooxygenase through the molecular docking studies.

Wei[71] has synthesized 125 curcuminoids and 14 heterocyclic curcuminoids and worked for evaluating the anti-inflammatory and antioxidant properties. The anti-inflammatory property was mediated by inhibition of monocyte chemoattractant protein-1, interleukin-6 (IL-6), interleukin-10 (IL-10), and tumor necrosis factor-alpha through cell-based NO assay. The anti-inflammatory evaluation showed that 73 compounds were successfully in inhibiting the NO production on IFN-ϒ/LPS stimulated RAW 264.7 microphages with IC_{50} range of 4.18–79.87 μM.

Novel curcuminoids with bis-acetamides have been prepared by Kirubavathi et al.[72] for enhancing their pharmacological activities. The synthesized entities have been examined for cytotoxicity, anti-inflammatory, and antioxidant activities in vitro. All the entities showed potent to good anti-inflammatory, cytotoxicity, and antioxidant activities with IC_{50} values in micromolar range.

Nieto et al.[73] have synthesized six new 3(5)-trifluoromethyl-5(3)-substituted-styryl-1*H*-pyrazoles and their tautomerism is studied both in solution and in solid state. Most of the compounds showed better inhibitory effect higher than 50% of the iNOS isoform; only two of the targeted compounds showed inhibition of about 50% with regards to the NOS inhibitory activity.

Five series of sulfonamides conjugated curcuminoids have been designed and prepared by Jaggi Lal et al.[74]. The prepared entities have been evaluated for in vitro antibacterial activity against Gram-(+) and Gram-(−) bacterial species viz. *Bacillus cereus*, *S. typhi*, *S. aureus*, *Escherichia coli*, and *Pseudomonas aeruginosa.* Antifungal activities of new entities have been marked against few pathogenic fungal species, viz., *Trichoderma viride*, *Aspergillus flavus*, *Curvularia*, and *Aspergillus niger* lunata. The cytotoxicity has been marked against human cell lines HeLa, Hep G2, QG-56, and HCT-116 by measuring IC_{50}. Some compounds displayed higher cytotoxicity than curcumin. The sulfonamides conjugated curcuminoids were also evaluated for anti-inflammatory activity in vivo.

A variety of new entities containing aromatic and heterocyclic aromatic curcuminoids have been synthesized by Khan et al.[75] The new compounds have been characterized and determined their anti-inflammatory activities by the animal model of female Wistar rats through oral administration. Among these, four novel curcuminoids notified as 1,7-bis(5-methylidenebutenolide)-1,6-heptadiene-3,5-dione, 1,7-bis(1-naphthyl)-1,6-hepta diene - 3,5-dione, 1,7-bis[2-(5-methylthiophenyl)]-1,6-heptadiene-3,5-dione, and 1,7-Bis[2-(5-methyfuranyl)]-1,6-heptadiene-3,5-dione in which the bis-methoxyphenyl group of curcumin was replaced with bis-dimethoxybutenolidyl, bis-furanyl, ascorbate, and bisnaphthyl derivatives. Of the curcuminoids, the furan 1,7-bis[2-(5-methyfuranyl)]-1,6-heptadiene-3,5-dione was the only compound to inhibit the production of TNF-α and IL-1β in a monocytic cell line THP-1 in vitro. The inactivity of 1,7-bis[2-(5-methyfuranyl)]-1,6-heptadiene-3,5-dione on the production of PGE_2 may be related to its absence of gastrotoxicity. This

1,7-bis[2-(5-methyfuranyl)]-1,6-heptadiene-3,5-dione may warrant the development of new low gastrotoxic anti-inflammatory agents with selective inhibitory activity of cytokine inflammatory mediators.

9.3.8 *RADIOPROTECTIVE EFFECT*

Srinivasan et al.[76] evaluated the radioprotective effect of curcuminoids analogs (bis-1,7-(2-hydrocyphenyl)-hepta-1,6-diene-3,5-dione) on gamma radiation-induced toxicity in cultures isolated from rat hepatocytes. The culture was treated with different concentrations of curcuminoids showed a significant decrease in the levels of those substances which react with thiobarbituric acid, DNA damage, and protected the primary culture hepatocytes against damage through gamma radiation by reducing peroxidation of membrane lipids and free radicals-induced DNA strand break formation. CUR-analog administration prior to radiation therapy can be useful for cancer patients to prevent normal cell damage.

Fukuda et al.[77] studied the radioprotective nature of newly synthesized curcuminoid against induced 11-Gy X-rays in experimental mice. The novel molecules prevented the intestinal mucosal damage in mice. Curcuminoids showed protective nature against gamma irradiation in the experimental mice during the radioprotection experiments. Molecules that are involved such as Bcl-2, p53, reactive carbonyl species (RCS), Bax, and cleaved caspase-3 were found to be related to radiation damage when tested immunohistochemically. New curcuminoids suppressed the accumulation of RCS and apoptosis-related molecules in irradiated cells.

9.4 CONCLUSION

The chapter comprehensively depicted the antitumor, antimalarial, antiproliferative, antioxidant, anti-inflammatory, cytotoxic, and radioprotective effects of derivatives and formulations of curcumin, a component of turmeric (*Curcuma longa*). The content will be highly beneficial to biologists, chemists, botanists, microbiologists, pharmacologists, zoologists, medicinal chemists, biotechnologists, pharmacognosist, and allied science students in deep learning, complete understanding, thoroughly inspiring, and further innovating new pharmacotherapeutically privileged synthesized hybrid derivatives, formulations, and marketed products in the near future.

KEYWORDS

- **turmeric**
- ***Curcuma longa***
- **pharmacological**
- **natural**
- **inhibitor**
- **synthesis**
- **formulations**
- **phytochemicals**

REFERENCES

1. Newman, D. J.; Cragg, G. M.; Snader, K. M. Natural Products as Sources of New Drugs over the Period. *J. Nat. Prod.* **2003,** *66*, 1022–1037.
2. Pandit, S.; Kim, H.; Kim, J.; Jeon, J. Separation of an Effective Fraction from Turmeric Against *Streptococcus mutans* Biofilms by the Comparison of Curcuminoid Content and Anti-acidogenic Activity. *Food Chem.* **2011,** *126*, 1565–1570.
3. Paramasivam, M.; Poi, R.; Banerjee, H.; Bandyopadhyay, A. High Performance Thin Layer Chromatographic Method for Quantitative Determination of Curcuminoids in *Curcuma longa* Germplasm. *Food Chem.* **2009,** *113*, 640–644.
4. Kçhler, F. E. *Kçhlers Medizinal-Pflanzen in Naturgetreuen Abbildungen Mit Kurz Erl_Uterndem Texte: Atlas Zur Pharmacopoea Germanica*; Gera-Untermhaus: Gera, 1887.
5. Ammon, H.; Wahl, M. A. Pharmacology of *Curcuma longa*. *Planta. Med.* **1991,** *57*, 1–7.
6. Aggarwal, B. B.; Sundaram, C.; Malani, N.; Ichikawa, H. Curcumin: The Indian Solid Gold. *Adv. Exp. Med. Biol.* **2007,** *595*, 1–75.
7. Thakur, R.; Puri, H. S.; Husain, A. *Major Medicinal Plants of India*; Central Institute of Medicinal and Aromatic Plants: Lucknow, 1989.
8. Tilak, J.; Banerjee, M.; Mohan, H.; Devasagayam, T. P. A. Antioxidant Availability of Turmeric in Relation to Its Medicinal and Culinary Uses. *Phytother. Res.* **2004,** *18*, 798–804.
9. Chaturvedi, T. P. Uses of Turmeric in Dentistry: An Update. *Indian J. Dent. Res.* **2009,** *20*, 107–109.

10. Mukerjee, A.; Vishwanatha, J. K. Formulation, Characterization and Evaluation of Curcumin-loaded PLGA Nanospheres for Cancer Therapy. *Anticancer Res.* **2009,** *29*, 3867–3875.
11. Perko, T.; Ravber, M.; Knez, Z.; Skerget, M. Isolation, Characterization and Formulation of Curcuminoids and In Vitro Release Study of the Encapsulated Particles. *J. Supercrit. Fluids.* **2015,** *103*, 48–54.
12. Kalpravidh, R. W.; Siritanaratkul, N.; Insain, P. Improvement in Oxidative Stress and Antioxidant Parameters in B-Thalassemia/Hb E Patients Treated with Curcuminoids. *Clin. Biochem.* **2010,** *43*, 424–429.
13. Changtam, C.; De Koning, H. P.; Ibrahim, H.; Sajid, M. S.; Gould, M. K.; Suksamrarn, A. Curcuminoid Analogs with Potent Activity Against Trypanosoma and Leishmania Species. *Eur. J. Med. Chem.* **2010,** *45*, 941–956.
14. Lim, H. S.; Park, S. H.; Ghafoor, K.; Hwang, S. Y.; Park, J. Quality and Antioxidant Properties of Bread Containing Turmeric (*Curcuma longa* L.) Cultivated in South Korea. *Food Chem.*, **2011,** *124*, 1577–1582.
15. Peret-Almeida, L.; Cherubino, A. P. F.; Alves, R. J.; Dufoss, E. L.; Gloria, M. B. A. Separation and Determination of the Physico-chemical Characteristics of Curcumin, Demethoxy Curcumin and Bisdemethoxycurcumin. *Food Res Int.* **2005,** *38*, 1039–1044.
16. Aditya, N. P.; Chimote, G.; Gunalan, K.; Banerjee, R.; Patankar, S.; Madhusudhan, B. Curcuminoids-loaded Liposomes in Combination with Arteether Protects Against Plasmodium Berghei Infection in Mice. *Exp. Parasitol.* **2012,** *131*, 292–299.
17. Khan, M. A.; El-Khatib, R.; Rainsford, K. D.; Whitehouse, M. W. Synthesis and Anti-inflammatory Properties of Some Aromatic and Heterocyclic Aromatic Curcuminoids. *Bioorg Chem.* **2012,** *40*, 30–38.
18. Yue, G. G. L.; Chan, B. C. L.; Hon, P. Immunostimulatory Activities of Polysaccharide Extract Isolated from *Curcuma longa. Int. J. Biol. Macromol.* **2010,** *47*, 342–347.
19. Tapal, A.; Tiku, P. K. Complexation of Curcumin with Soy Protein Isolate and Its Implications on Solubility and Stability of Curcumin. *Food Chem.*, **2012,** *130*, 960–965.
20. Panahi, Y.; Saadat, A.; Beiraghdar, F.; Nouzari, S. M. H.; Jalalian, H. R.; Sahebkar, A. Antioxidant Effects of Bioavailability-enhanced Curcuminoids in Patients with Solid Tumors: A Randomized Double-Blind Placebo-controlled Trial. *J. Funct. Foods.* **2014,** *6*, 615–622.
21. Zhan, P. Y.; Zeng, X. H.; Zhang, H. M.; Li, H. H. High-efficient Column Chromatographic Extraction of Curcumin from *Curcuma longa. Food Chem.* **2011,** *129*, 700–703.
22. Jalili-Nik, M.; Soltani, A.; Moussavi, S.; Ghayour-Mobarhan, M.; Ferns, G. A.; Hassanian, S. M.; Avan, A. Current Status and Future Prospective of Curcumin as a Potential Therapeutic Agent in the Treatment of Colorectal Cancer. *J. Cell Physiol.* **2018,** *233*, 6337–6345.
23. Qiu, P.; Zhanga, S.; Zhou, Y.; Zhua, M.; Kang, Y.; Chena, D.; Wang, J.; Zhou, P.; Lia, W.; Xu, Q.; Jin, R.; Wu, J.; Liang, G. Synthesis and Evaluation of Asymmetric Curcuminoid Analogs as Potential Anticancer Agents That Downregulate NF-Kb Activation and Enhance the Sensitivity of Gastric Cancer Cell Lines to Irinotecan

Chemotherapy. *Euro. J. Med. Chem.* (accepted). DOI: 10.1016/J.Ejmech. 2017. 08.022.

24. Gyuris, M.; Hackler Jr. L.; Nagy, L.; Alfoldi, R.; Redei, E.; Marton, A.; Vellai, T.; Farago, N.; Ozsvari, B.; Hetenyi, A.; Toth, G.; Sipos, P.; Kanizsai, I.; Puskas, L. Mannich Curcuminoids as Potent Anticancer Agents. *Arch. Pharm. Chem. Life Sci.* **2017,** *350*, 1700005.
25. Yodkeeree, S.; Chaiwangyen, W.; Garbisa, S.; Limtrakul, P. Curcumin, Demethoxy Curcumin and Bisdemethoxycurcumin Differentially Inhibit Cancer Cell Invasion Through the Down-Regulation of MMPS and UPA. *J. Nutr. Biochem.* **2009,** *20*, 87–95.
26. Ali, N. M.; Yeap, S. K.; Abu, N.; Lam Lim K.; Ky, H.; Mat Pauzi, A. Z.; Yong Ho, W.; Tan, S.; Alan-Ong, H.; Zareen, S.; Alitheen, N.; Akhtar, N. Synthetic Curcumin Derivative DK1 Possessed G2/M Arrest and Induced Apoptosis Through Accumulation of Intracellular ROS in MCF-7 Breast Cancer Cells. *Cancer Cell Int.* **2017,** *17*, 30. DOI: 10.1186/S12935-017-0400-3.
27. Gurung, R.; Gong, S.; Dhakal, D.; Thi Le, T.; Jung, N.; Jung, H.; Jin Oh, T.; Sohng, J. K. Synthesis of Curcumin Glycosides with Enhanced Anticancer Properties Using One-pot Multienzyme Glycosylation Technique. *J. Microbiol. Biotechnol.* **2017,** *27* (9), 1639–1648.
28. Basile, V.; Ferrari, E.; Lazzari, S.; Belluti, S.; Pignedoli, F.; Imbriano, C. Curcumin Derivatives: Molecular Basis of Their Anti-cancer Activity. *Biochem. Pharmacol.* **2009,** *78*, 1305–1315.
29. Ahsan, M. J. Evaluation of Anticancer Activity of Curcumin Analogues Bearing a Heterocyclic Nucleus. *Asian Pac. J. Cancer Prev.* **2016,** *17*, 1739–1744.
30. Prasad, C. P.; Rath, G.; Mathur, S.; Bhatnagar, D.; Ralhan, R. Expression Analysis of MASPIN in Invasive Ductal Carcinoma of Breast and Modulation of Its Expression by Curcumin in Breast Cancer Cell Lines. *Chem. Biol. Interact.* **2010,** *183*, 455–461.
31. Anthwal, A.; Thakur, B.; Rawat, M. S. M.; Rawat, D. S.; Tyagi, A.; Aggarwal, B. Synthesis, Characterization and *In Vitro* Anticancer Activity of C-5 Curcumin Analogues with Potential to Inhibit TNF-5-Induced NF-5B Activation. *Biomed. Res. Int.* **2014,** 524161. http://dx.doi.org/10.1155/ 2014/524161.
32. Einbond, L. S.; Wu, H.; Kashiwazaki, R.; et al. Carnosic Acid Inhibits the Growth of ER-Negative Human Breast Cancer Cells and Synergizes with Curcumin. *Fitoterapia* 2012, *83*, 1160–1168.
33. Lin, H.; Lin, J.; Ma, J.; et al. Demethoxycurcumin Induces Autophagic and Apoptotic Responses on Breast Cancer Cells in Photodynamic Therapy. *J. Funct. Foods.* **2015,** *12*, 439–449.
34. Balaji, B.; Balakrishnan, B.; Perumalla, S.; Karande, A.; Chakravarty, A. A Photoactivated Cytotoxicity of Ferrocenyl-Terpyridine Oxovanadium(IV) Complexes of Curcuminoids. *Eur. J. Med. Chem.* **2014,** *85*, 458–467.
35. Ye, M.; Zhao, Y.; Li, Y. Curcumin Reverses Cisplatin Resistance and Promotes Human Lung Adenocarcinoma A549/DDP Cell Apoptosis Through HIF-1a and Caspase-3 Mechanisms. *Phytomedicine* **2012,** *19*, 779–787.
36. Feng, L.; Li, Y.; Song, Z.; Li, H.; Huai, Q. Synthesis and Biological Evaluation of Curcuminoid Derivatives. *Chem. Pharm. Bull.* **2015,** *63*, 873–881.

37. Rao, A.; Prasad, E.; Seelam, D.; Ansari, I. Synthesis and Biological Evaluation of Glucosyl Curcuminoids. *Arch. Pharm. Chem. Life Sci.* **2014,** *347*, 1–6.
38. Cao, Y.; Jing Li, H.; Song, Z.; Li, Y.; Huai, Q. Synthesis and Biological Evaluation of Novel Curcuminoid Derivatives. *Molecules* **2014,** *19*, 16349–16372. DOI: 10.3390/ Molecules191016349.
39. Puneeth, H. R.; Ananda, H.; Kumar, K. S. S.; Rangappa, K. S.; Sharada, A. C. Synthesis and Antiproliferative Studies of Curcumin Pyrazole Derivatives. *Med. Chem. Res.* **2016,** *25*, 1842–1851. DOI: 10.1007/S00044-016-1628-5.
40. Ahsan, M. J.; Khalilullah, H.; Yasmin, S.; Jadav, S.; Govindasamy, J. Synthesis, Characterisation, and *In Vitro* Anticancer Activity of Curcumin Analogues Bearing Pyrazole/Pyrimidine Ring Targeting EGFR Tyrosine Kinase. *Bio. Med. Res. Int.* **2013**, 239354. http://dx.doi.org/10.1155/ 2013/239354.
41. Lozada-García, M.; Enríquez, R.; Ramírez-Apán, T.; Nieto-Camacho, A.; Palacios-Espinosa, F.; Custodio-Galván, Z.; Soria-Arteche, O.; Pérez-Villanueva, J. Synthesis of Curcuminoids and Evaluation of Their Cytotoxic and Antioxidant Properties. *Molecules* **2017,** *22*, 633. DOI: 10.3390/Molecules22040633.
42. Zheng, Q.; Yang, Z.; Yu, L.; Yan Ren, Y.; Huang, Q.; Liu, Q.; Ma, X.; Chen, Z.; Wang, Z.; Zheng, X. Synthesis and Antioxidant Activity of Curcumin Analogs. *J. Asian Nat. Prod. Res.* **2016**. DOI: 10.1080/10286020.2016.1235562.
43. Sahu, P. K. Design, Structure Activity Relationship, Cytotoxicity and Evaluation of Antioxidant Activity of Curcumin Derivatives/Analogues. *Eur. J. Med. Chem.* **2016**. DOI: 10.1016/J.Ejmech.2016.05.037.
44. Mošovskáa, S.; Petáková, P.; Kaliňák, M.; Mikulajovác, A. Antioxidant Properties of Curcuminoids Isolated from *Curcuma longa* L. *Acta Chim. Slovaca* **2016,** *9* (2), 130–135. DOI: 10.1515/Acs-2016-0022.
45. Sokmen, M.; Khan, M. A. The Antioxidant Activity of Some Curcuminoids and Chalcones. *Inflammopharmacol.* **2016,** *24*, 81–86. DOI: 10.1007/S10787-016-0264-5.
46. Sahu, P. K.; Sahu, P. K.; Sahu, P. L.; Agarwal, D. D. Structure Activity Relationship, Cytotoxicity and Evaluation of Antioxidant Activity of Curcumin Derivatives. *Bioorg. Med. Chem. Lett.* **2016,** *26*, 1342–1347.
47. Li, Q.; Chen, J.; Luo, S.; Xu, J.; Huang, Q.; Liu, T. Synthesis and Assessment of the Antioxidant and Antitumour Properties of Asymmetric Curcumin Analogues. *Eur. J. Med. Chem.*, **2015**. DOI: 10.1016/J.Ejmech.2015.02.005.
48. Bandgar, B.; Jalde, S.; Korbad, B.; Patil, S.; Chavan, H.; Kinkar, S.; Adsul, L.; Shringare, S.; Nile, S. Synthesis and Antioxidant, Cytotoxicity and Antimicrobial Activities of Novel Curcumin Mimics. *J. Enzyme Inhib. Med. Chem.* **2012,** *27* (2), 267–274. DOI: 10.3109/ 14756366.2011.587416.
49. Li, P.; Liu Z. Ferrocenyl-substituted Curcumin: Can It Influence Antioxidant Ability to Protect DNA. *Eur. J. Med. Chem.* **2011,** *46*, 1821–1826.
50. Parvathy, K. S.; Negi, P. S.; Srinivas, P. Curcumineamino Acid Conjugates: Synthesis, Antioxidant and Antimutagenic Attributes. *Food Chem.* **2010,** *120*, 523–530.
51. Roos, F.; Binder, K.; Rutz, J.; Maxeiner, S.; Bernd, A.; Kippenberger, S.; Zöller, N.; Chun, F. K.-H.; Juengel, E.; Blaheta, R. A. The Antitumor Effect of Curcumin in Urothelial Cancer Cells Is Enhanced by Light Exposure In Vitro. *Evid.-Based Complementary Altern. Med.* **2019,** 6374940. https://doi.org/10.1155/ 2019/6374940.

52. Hintzpeter, J.; Hornung, J.; Ebert, B.; Martin, H.; Maser E. Curcumin Is a Tight Binding Inhibitor of the Most Efficient Human Daunorubicin Reductase—Carbonyl Reductase. *Chem. Biol. Interact.* **2015,** *234*, 162–168.
53. Jiang, J.; Jin, X.; Zhang, H.; Su, X.; Qiao, B.; Yuan Y. Identification of Antitumor Constituents in Curcuminoids from *Curcuma longa* L. Based on the Composition-Activity Relationship. *J. Pharm. Biomed. Anal.* **2012,** *70*, 664–670.
54. Dahan, A. Zeyad.; Hussain, S.; Gyananath, G.; Zubaidha, P. K. Aldose Reductase Inhibitory Activity Studies of THERACURMIN. *Nat. Prod. Res.* **2018,** *32* (18), 2248–2251.
55. Kondhare, D. D.; Gyananath, G.; Tamboli, Y.; Kumbhar, S. S.; Choudhari, P. B.; Bhatia, M. S.; Zubaidha, P. K. An Efficient Synthesis of Flavanones and Their Docking Studies with Aldose Reductase. *Med. Chem. Res.* **2017,** *26*, 987–998.
56. Ghorbani, Z.; Hekmatdoost, A.; Mirmiran, P. Anti-hyperglycemic and Insulin Sensitizer Effects of Turmeric and Its Principle Constituent Curcumin. *Int. J. Endocrinol. Metab.* **2014,** *12* (4), 18081.
57. Yuan, X.; Li, H.; Bai, H., Su, Z.; Xiang, Q.; Wang, C.; Zhao, B.; Zhang, Y.; Zhang, Q.; Chu, Y.; Huang, Y. Synthesis of Novel Curcumin Analogues for Inhibition of 11β-Hydroxysteroid Dehydrogenase Type 1 with Anti-diabetic Properties. *Eur. J. Med. Chem.*, **2014,** *77*, 223–230.
58. Kim, T.; Davis, J.; Zhang, A.; He, X.; Mathews, S. Curcumin Activates AMPK and Suppresses Gluconeogenic Gene Expression in Hepatoma Cells. *Biochem. Biophys. Res. Commun.* **2009,** *388*, 377–382.
59. Dohutia, C.; Chetia, D.; Gogoi, K.; Sarma, K. Design, *In Silico* and *In Vitro* Evaluation of Curcumin Analogues Against *Plasmodium falciparum*. *Exp. Parasitol.* **2017**. DOI: 10.1016/J.Exppara.2017.02.006.
60. Dohutia, C.; Chetia, D.; Gogoi, K.; Sarma, K.; Bhattacharyya, D. Molecular Docking, Synthesis and *In Vitro* Antimalarial Evaluation of Certain Novel Curcumin Analogues. *Braz. J. Pharm. Sci.* **2017,** *53* (4), E00084.
61. Balaji, S. N.; Ahsan, J.; Jadav, S.; Trivedi, V. Molecular Modelling, Synthesis, and Antimalarial Potentials of Curcumin Analogues Containing Heterocyclic Ring. *Arabian J. Chem.* **2015**. http://dx.doi.org/10.1016/J.Arabjc.2015.04.011.
62. Manohar, S.; Khan, S.; Kandi, S.; Raj, K.; Sun, G.; Yang, X.; Molina, A.; Ni, N.; Wang, B.; Rawat, D. Synthesis, Antimalarial Activity and Cytotoxic Potential of New Monocarbonyl Analogues of Curcumin. *Bioorg. Med. Chem. Lett.* **2013,** *23*, 112–116.
63. Shukla, A.; Singh, A.; Singh, A.; Pathak, L. P.; Shrivastava, N.; Tripathi, P. K.; Singh, M. P.; Singh, K. Inhibition of *P. Falciparum* PFATP6 by Curcumin and Its Derivatives: A Bioinformatic Study. *Cell. Mol. Biol.* **2012,** *58* (1), 182–186.
64. Chen, C.; Long, L.; Zhang, F.; Chen, Q.; Chen, C.; Yu, X.; Liu, Q.; Bao, J.; Long, Z. Antifungal Activity, Main Active Components and Mechanism of *Curcuma longa* Extract Against *Fusarium graminearum*. *PLoS One* **2018,** *13* (3), E0194284. https://doi.org/10.1371/ Journal.Pone.0194284.
65. Caprioglio, D.; Torretta, S.; Ferrari, M.; et al. Triazole-Curcuminoids: A New Class of Derivatives for 'Tuning' Curcumin Bioactivities. *Bioorg. Med. Chem.* **2016,** *24*, 140–152.

66. Shi, Q.; Wada, K.; Ohkoshi, E.; et al. Antitumor Agents 290. Design, Synthesis, and Biological Evaluation of New LNCaP and PC-3 Cytotoxic Curcumin Analogs Conjugated with Anti-androgens. *Bioorg. Med. Chem.* **2012,** *20*, 4020–4031.
67. Zhang, Q.; Zhong, Y.; Yan, L.; Sun, X.; Gong, T.; Zhang, Z. Synthesis and Preliminary Evaluation of Curcumin Analogues as Cytotoxic Agents. *Bioorg. Med. Chem. Lett.* **2011,** *21*, 1010–1014.
68. Ferrari, E.; Lazzari, S.; Marverti, G.; Pignedoli, F.; Spagnolo, F.; Saladini, M. Synthesis, Cytotoxic and Combined cDDP Activity of New Stable Curcumin Derivatives. *Bioorg. Med. Chem.* **2009,** *17*, 3043–3052.
69. Jadhava, S. Y.; Bhosalea, R. B.; Shiramea, S. P.; Patilb, S. B.; Kulkarnic, S. D. PEG Mediated Synthesis and Biological Evaluation of Asymmetrical Pyrazole Curcumin Analogues as Potential Analgesic, Anti-inflammatory and Antioxidant Agents. *Chem. Biol. Drug Des.* **2015,** *85* (3), 377–384. DOI: 10.1111/Cbdd.12416.
70. Banuppriya, G.; Sribalan, R.; Padmini, V.; Shanmugaiah V. Biological Evaluation and Molecular Docking Studies of New Curcuminoid Derivatives: Synthesis and Characterization. *Bioorg. Med. Chem. Lett.* **2016,** *26*, 1655–1659.
71. Leong, S. W. Design, Synthesis, Anti-Inflammatory Activity Evaluation and Structure-Activity Relationship (SAR) Study of Diarylpentanoids Derivatives. Ph.D. Thesis, University of Putra Malaysia, 2014.
72. Sribalan, R.; Kirubavathi, M.; Banuppriya, G.; Padmini, V. Synthesis and Biological Evaluation of New Symmetric Curcumin Derivatives. *Bioorg. Med. Chem. Lett.* **2015,** *25*, 4282–4286.
73. Nieto, C. I.; Cabildo, M. P.; Cornago, M. P.; et al. Synthesis, Structure and Biological Activity of 3(5)-Trifluoromethyl-1H-Pyrazoles Derived from Hemicurcuminoids. *J. Mol. Struct.* **2015,** *1100*, 518–529.
74. Lal, J.; Gupta, S.; Thavaselvam, D.; Agarwal, D. Biological Activity, Design, Synthesis and Structure Activity Relationship of Some Novel Derivatives of Curcumin Containing Sulphonamides. *Eur. J. Med. Chem.* **2013,** *64*, 579–588.
75. Khan, M. A.; El-Khatib, R.; Rainsford, K. D.; Whitehouse, M. W. Synthesis and Anti-inflammatory Properties of Some Aromatic and Heterocyclic Aromatic Curcuminoids. *Bioorg. Chem.* **2012,** *40*, 30–38.
76. El-Gazzar, M. G.; Zaher, N. H.; El-Hossary, E. M.; Ismail, A. F. M. Radio-protective Effect of Some New Curcumin Analogues. *J. Photochem. Photobiol. B* **2016**. DOI: 10.1016/ J.Jphotobiol. 2016. 08.002.
77. Koji, F.; Uehara, Y.; Nakata, E.; Inoue, M.; Iwabuchi, Y.; Shibata, H. A Diarylpentanoid Curcumin Analog Exhibits Improved Radioprotective Potential in the Intestinal Mucosa. *Int. J. Radiat. Bio.* **2016,** *92* (7), 388–394.

CHAPTER 10

Emerging Highlights on Natural Prodrug Molecules with Multifarious Therapeutic Perspectives

MOJABIR HUSSEN ANSARI[1], VAIBHAV SHENDE[2], and DEBARSHI KAR MAHAPATRA[3*]

[1]*Department of Quality Assurance, Gurunanak College of Pharmacy and Technical Institute, Nagpur 440026, India*

[2]*Department of Pharmaceutics, Gurunanak College of Pharmacy and Technical Institute, Nagpur 440026, India*

[3]*Department of Pharmaceutical Chemistry, Dadasaheb Balpande College of Pharmacy, Nagpur 440037, India*

[*]*Corresponding author. E-mail: mahapatradebarshi@gmail.com*

ABSTRACT

Natural prodrugs are the chemical substances or compounds obtained from living species like plants, animals, microorganisms, and marine sources. Modern-day natural products in practical applications are small molecules modified with the aid of chemical synthesis and utilized for therapy, fortified food, the element of drug discovery, and for various healing purposes. Most of the medication for diverse sicknesses metabolizes rapidly by the process of first-pass effect that consequently results in drug inactivation, and formation of harmful or poisonous metabolites in the human body. Most of the prodrugs obtained from natural sources like plants, animals, and marine resources, specifically undergo chemical transformation into non-toxic compounds, in addition to prevention from complete drug inactivation. The chapter highlights some of the updates on prodrugs obtained from natural products.

10.1 INTRODUCTION

Prodrugs are defined as the inactive form of their parent drug. However, there are some prodrugs that might be active before undergoing enzymatic or chemical interconversion themselves.[1] Prodrugs were first brought in 1958 to indicate a pharmacologically inactive chemical moiety that may be enforced to regulate the physicochemical residences of a drug to build its effectiveness and reduce its related toxicity effects.[2] Nature always stands very first in the treatment of various ailments by providing armamentarium from the plant, mineral source, animal, and microbial species to treat various kinds of diseases. The traditional system of medicine like Ayurveda, Unani, Chinese, and Siddha system incorporates information on natural products for about 100 decades.[3] The synthetic drugs cause harmful effects on the body by weakening the immune system or certain damaging to the internal organs and precipitate diseases like gastrointestinal bleeding, ulcer, increasing threat of coronary heart disorder, and fluid retention. Natural products boost the immunity of the body to fight back the diseases.[4] The secondary metabolites obtained from natural products play an important role in the development of a large number of drugs. Prodrugs found naturally in numerous botanical phytoconstituents or derived from the synthetic or semi-synthetic process during rational drug design or unintentionally during drug development have their own importance.[5] Natural prodrugs found unintentionally during the drug development process such as heroin, codeine, irinotecan, aspirin, psilocybin, and several antiviral nucleosides that are into modern applications.[6]

10.1.1 INDIVIDUAL NATURAL PRODRUGS

10.1.1.1 PRONTOSIL

Prontosil, also known as Prontosil rubrum, is a dark red color dye. It was initially given orally as a hydrochloride (HCl) salt but later preferred as a free base as it was less staining. Moreover, this dye is very insoluble in aqueous media and their recourse becomes extremely difficult. Over time, it was formulated to make it more water soluble. Recently, a library of organized, patented compounds has been seen for fabricating derivatives with a greater solubility and can be used in formulating injectables.

Prontosil soluble (originally Streptozon S) came into practice through the chemical reaction among the diazotized sulphanilamide with the 2-acetamino-8-hydroxynaphthalene-3,6-disulphonic acid.[7] Prontosil later changed into a prodrug form that is specifically used in the formation and improvisation of the sulfonamide molecules. In mixture with diaminopyrimidines, they inhibit the folate pathway in bacteria and are used widely for treating bacteria-oriented infections. In olden times, diaminopyrimidines were used along with sulphonamides besides for special indications (infectious diseases) or for the remedy of cystitis.[8]

10.1.2 BUTYRIN

Butyrin (tributyrin), the prodrug of butyric acid, is an essential triglyceride molecule that is naturally present in the butter and also in some butter products. It is a fatty liquid with acrid flavor and is specifically made up of ester of butyric acid and glycerol. Butyrate is also a product achieved by the fermentation process that took place mainly in the distal colon by the microorganism.[9] Butyric acid and its derivatives possess notable antimicrobial activities at different concentrations against *Salmonella typhimurium* and *Clostridium perfringens*. Butyric acid, when administered alone as a therapeutic agent often shows compromised pharmacological activity (particularly, short half-life) as a result of rapid metabolism.[10] In addition to it, the strong odor of butyrate hinders patient compliance; therefore, direct oral consumption of butyric acid is unacceptable. Although in a study, it was found that the best antimicrobial inhibition was reported by butyric acid and its derivative without the addition of lipase.[11] Although focusing on veterinarian applications, recently, it has been discovered that the glycerides and individual administration of butyrin in broiler chickens feed concurrently extends the carcass weight and breast meat. Butyrate improves the epithelization process and brings about collagen lysis by reducing the matrix metalloproteinase release.[12]

10.1.3 MELATONIN

Melatonin (5-methoxy-N-acetyltryptamine) was first discovered by Aaron Lerner in the year 1958 from bovine pineal. Retina, platelets, bone marrow cells, lymphocytes, skin, Harderian gland, gastrointestinal tract, cerebellum, etc., are also reported as a source of extrapineal supply of melatonin. Melatonin is mainly prepared by the pinealocytes from an essential amino acid tryptophan. Tryptophan is synthesized by hydroxylation of 5-hydroxytryptophan by the enzyme tryptophan-5-hydroxylase then further changed into decarboxylated form with the aid of the 5-hydroxytryptophan decarboxylase enzyme. The synthesis and secretion of melatonin are enhanced by the way of darkness and inhibited by the way of light.[13] Melatonin is an omnipresent molecule having natural and powerful antioxidant proprieties and it is safe when administered exogenously. Melatonin also possesses potential anti-inflammatory effects due to inhibiting inflammasome activation. Melatonin also owns its anti-apoptotic activities particularly by inhibiting Caspase-3 cleavage and mPTP opening. The presence and the inter-distance of 5-methoxy group and the N-acetyl chain are the two critical factors in the determination of specificity and amphiphilicity of melatonin.[14] Scientific studies on MT_1 and MT_2 receptor knockout mice have resulted that MT_1 and MT_2 receptors play specific roles in sleep. Melatonin and melatonin agonists also play essential roles in the remedy of insomnia by activating MT_1 and MT_2 melatonin receptors.[15]

10.1.4 ROMIDEPSIN

Romidepsin is a bicyclic depsipeptide that was first isolated from a Gram-negative rod-shaped single polar flagellum bacteria *Chromobacterium violaceum.*[16] In early 1990s, romidepsin remained an important fermentation product for treating tumor (histone deacetylase class-I inhibitor; however, the mechanism is not fully known) forms such as glioblastoma, leukemia, lymphoma, myeloma, and breast, colorectal, gastrointestinal, lung, ovarian, pancreatic, and prostate cancer and also for treating the microbial infections.[17] It is a prodrug that requires a reduction of its disulfide bonds to activate its less stable form. Histone acetyltransferases and histone deacetylases control histone acetylation by the way of direct addition of acetyl groups to the lysine residues within the amino-terminal histone tails, which neutralizes the part of the protein and relaxes the chromatin structure.[18] Romidepsin is also classified as an epigenetic agent that introduces stable genetic changes by interfering with the gene expression and their function, without any corresponding changes in the

DNA sequence.[19] In recent years, small molecules of histone deacetylase (HDAC) inhibitors have owned the position of strong anticancer agents, many of which are now FDA approved anticancer agents, which have challenged the position of romidepsin.[20]

10.1.5 BAICALIN

Baicalin is a prodrug flavonoid of baicalein, which has been acquired from the traditional Chinese medicinal plant *Scutellaria Baicalensis Georgi*. It shows many crucial activities such as antitumor, anti-inflammatory, anti-bacterial, immunostimulant, anti-allergic, and antiviral effects. Baicalein is more effective compared to baicalin because it is absorbed more slowly and to a lesser quantity than baicalein. However, it is very difficult to remove baicalein without delay because of its low content (0.2–0.5%) in the extract. A higher and reliable method of obtaining baicalein is by the enzyme hydrolysis from baicalin, which is present in better content material in the extract (6–10%).[21,22] Cadmium-induced hepatic cytotoxicity, oxidative stress, and histomorphometric changes are additionally prevented by means of baicalin.[23] Baicalin has also been investigated to have 5α-reductase inhibiting activity that is useful as a hair growth stimulant and is presently commercially available in several brands of shampoo.[24] Based on the molecular studies, it is proved that the anti-inflammatory activity occurred due to its bioactive chemical flavones. Along with anti-inflammatory activities, the flavones isolated from *Scutellarin* have also shown cytostatic and cytotoxic properties against many human cancer cell lines.[25] Baicalin also inhibited the growth of *Chlamydia trachomatis*, a prevalent sexually transmitted bacterial pathogen in humans that causes pelvic inflammatory disease, ectopic pregnancy, and infertility in women. Baicalin also protects the human fibroblasts toward ultraviolet B-triggered cyclobutane pyrimidine dimers formation.[26,27]

10.1.6 PHYTOESTROGEN

Phytoestrogens are compounds that have steroidal estrogen-like biological activity, and are primarily extracted from more than 300 plant species (several edible and/or medicinal plants, particularly Leguminosae family). It is a susceptible class of estrogens, which are extracted from the flowers (e.g., soy, red clover, kudzu, hops, liquorice, rhubarb, yam, and chaste-berry) or derived from plant precursors and are utilized by animals or humans. Phytoestrogens are widely present in fruits, vegetables, and complete grains that are commonly consumed by humans. A single plant also may contain multiple classes of phytoestrogen. For example, the soybean is rich in isoflavones, whereas the soy sprout is wealthy in coumestrol, the primary coumestan.[28–31] Several groups of chemical compounds such as isoflavones (daidzein and genistein), coumestans, and lignans, which are structurally similar to the endogenous estrogens, are also present but these compounds show both estrogenic and anti-estrogenic effects. Looking insight to it, daidzein can be derived from formononetin and within the intestine further metabolized into desmethylangolensin (DMA).[32–33] Phytoestrogens administered to women or given in diets have a low incidence of estrogen-associated cancers, cardiovascular diseases, and climacteric symptoms. Phytoestrogens in the regular human food regimen encompass several substances that can be still unidentified.[34]

10.1.7 SENNOSIDE A

Senna (or Tinnevelly senna), obtained from *Cassia angustifolia* Vahl (Family: Fabaceae), is a well-known herbal product since ancient time due to laxative and purgative action (due to the presence of anthraquinone glycosides; sennoside A and sennoside B).[35,36] Apart from its well-known laxative properties, sennoside A is also used in the treatment of splenic enlargement, tumor, jaundice, diabetes, bronchitis, and leprosy. The plant is also recommended to grow in the wastelands, as it is eco-friendly and does require less and frequent irrigation. It can provide permanent green cover in the arid areas as the plant is perennial and easily established.[37-39]

10.1.8 BARBALOIN

Barbaloin (10-β-D-glucopyranosyl-1,8-dihydroxy-3-hydroxymethyl-9-(10*H*)-anthracenore) is a main bioactive compound in Aloe Vera (Liliaceous plant family). It is a cactus-like plant that grows in hot, dry climates.[40] The word "Aloe" derives from the Arabic word "Alloeh" meaning shining sour substance while "Vera" in Latin means true. There are more than 300 species of Aloe plant, but *Aloe barbadensis* species has the greatest medicinal properties.[41]. Barbaloin, also known as aloin, has diverse pharmacological effects that include anti-inflammatory, anti-oxidant, antitumor, purgative, and antiprotozoal effects.[42]. The purgative movement of barbaloin is triggered with the aid of *Eubacterium sp.*, which converts barbaloin to aloe-emodin anthrone. Barbaloin inhibits the rat colonic Na^+-K^+-ATPase in vitro and improved paracellular permeability across the rat colonic mucosa in vivo.. Barbaloin pretreatment attenuates the myocardial ischemia–reperfusion injury via activation of AMP-activated protein kinase (AMPK) and restores infarction size.[44].

10.1.9 GLYCYRRHIZIN

Glycyrrhizin, the principal constituent of liquorice, has steroidal, anti-inflammatory, antitumor, antiallergic, hepatoprotective, antihepatitis, antichronic bronchitis, antigastritis, anti-immunological, and antiviral activities.[45] It is a widely used as oral sweetener, oriental medicine, absorption enhancer (nasal and rectal absorption enhancement of antibiotic, insulin, and calcitonin), and transdermal enhancer of drugs.[46] It contains two glucuronosyl moieties linked to the steroids.[47] The compound is hydrolyzed by a stepwise manner to monoglucuronide and then into glycyrrhetinic acid by the help of endogenous and β-glucuronidases present in the intestinal lumen.[48] Glycyrrhizin is a glycosylated saponin containing one molecule of glycyrretinic acid with structural similarities to hydrocortisone, and two molecules of glucuronic acid.[49] However, the administration of glycyrrhizin may cause pseudo-hypercorticosteroidism, along with sodium and water retention, high blood pressure, and hypokalemia.[50]

10.1.10 LOVASTATIN

Lovastatin (also known as mevinolin) is the primary drug of a brand-new class of cholesterol-reducing drugs that competitively inhibit the 3-hydroxy-3-methylglutaryl coenzyme A (HMG-CoA) reductase, a rate-limiting enzyme within the biosynthesis of cholesterol.[51]. Lovastatin, a fungal metabolite was originally isolated from cultures of *Monascus ruber* and *Aspergillus terreus.*[52] Inhibition of hepatic cholesterol synthesis with lovastatin is achieved by a large (2–70-fold) increase in HMG-CoA reductase activity ex vivo in the rat. Lovastatin in combination with cholestyramine in the diet of rats produces a greater increase in HMG-CoA reductase activity than with either drug alone.[53] HMG-CoA reductase inhibitors also block the synthesis of cholesterol which has been reported to induce apoptosis in the prostate cancer cells. The exact mechanism(s) of lovastatin prompted apoptosis is not clear, but it can be resulting from the deficiency of some cholesterol pathway metabolites which include geranyl and farnesyl pyrophosphate.[54]

10.1.11 RESVERATROL

Resveratrol (trans-3, 5, 4-trihydroxystilbene), a plant-derived polyphenolic phytoalexin found in grapes, wine, peanuts, and cranberries has been said to have anticarcinogenic, antioxidative, phytoestrogen, platelet aggregation, and coagulation, modifies eicosanoid synthesis, modulate lipoprotein mechanisms, and cardioprotective activities.[55,56] Resveratrol has also proven to increase the lifespan of evolutionarily distant species which include *S. cerevisiae, C. elegans*, and *D. melanogaster* in a Sir2-dependent manner.[57] The scientific study reported that trans-resveratrol has health benefits including anticarcinogenic effects and protection against cardiovascular disease as well as in Alzheimer's disease.[58,59] The phytochemical additionally play a crucial role in the cellular response by modulating the enzymes concerned with the stress response, which include quinone reductase 2 (QR2), a cytosolic enzyme that facilitates the production of damage-activated quinone and reactive oxygen species (ROS).[60] Resveratrol is like a phytoestrogen and has some structural similarity to diethylstilboestrol. But, it has a more affinity for the estrogen receptor-β (ER-β) than α and transcriptionally activates ER-β at small concentrations.[61,62]

10.1.12 URSOLIC ACID

Ursolic acid is a pentacyclic triterpenoid used as an anti-inflammatory, anti-invasion, and anticancer agent (skin-tumor prevention, breast, leukemia, prostate, lung, melanoma, and endometrial cancer cells).[63,64] It is structurally very alike to asiatic acid and it broadly exists in all components of plant life and has been pronounced to possess a variety of biological effects.[65] Ursolic acid has actions on the growth of collagen content in human pores and skin.[66] However, the molecular objectives and mechanism(s) underlying ursolic acid are not completely characterized, and the effect in acute stroke is still unknown.[67,68]

10.1.13 HESPERIDIN

Hesperidin (vitamin P) is a β-7-rutinoside of hesperitin, and is a member of the flavanone glycoside of flavonoids. Hesperidin is a prodrug of hesperitin, which consists of an aglycone and a disaccharide, rutinose. It is isolated in massive amounts as inexpensive by-product from the rinds of a few citrus species (sweet orange and lemon) such as *Citrus aurantium, Citrus sinensis,* and *Citrus unshiu.*[69] The phytochemical demonstrates antiallergenic, anticarcinogenic, antihypotensive, antimicrobial, antidiabetic, antiulcer, anti-inflammatory, antimicrobial, antiallergic, and vasodilator properties.[70,71] The hypoglycemic effects of hesperidin are partly mediated through hepatic glucose-regulating enzymes in C57BL/KsJ-db/db Mice.[72,73] Hesperidin is effectively utilized as a supplemental agent in the remedy protocols of complementary settings.[74] Various researches have focused on the potential use of hesperidin as free radical scavengers and inhibitors of lipid peroxidation to prevent oxidative damage.[75,76]

10.1.14 ARBUTIN

Arbutin (4-hydroxyphenyl-β- D-glucopyranoside) is a hydroquinone derivative also known as β-arbutin.[77] This simple phenol glucoside can be extracted from plants and can also be biosynthesized especially by *Ericaceae* (silvan undershrub species) and *Saxifragaceae* family.[78] It is a famous tyrosinase inhibitor (inhibitory effect on the hydroxylation reaction of tyrosinase) that has an application in skin whitening.[79] Substituted arbutin derivatives such as 6-*p*-coumaryl, 6-cafferyl, and 6-*p*-hydroxybenzoyl arbutin have also been isolated from plants.[80] In erythrocytes, arbutin possess long-lasting radical scavenging properties and also protects the membrane lipid from oxidative stress in human skin fibroblasts.[81] Natural sources of these plants are limited and these species are often protected in European countries.[82] Recently, arbutin was found to be effective in post-inflammatory hyperpigmentation (PIH), which is a reactive hypermelanosis and sequela of most inflammatory skin conditions.[83]

10.1.15 L-DOPA

L-DOPA (3,4-dihydroxyphenylalanine) is a prodrug of dopamine. It is a fascinating compound that plays important roles in the field of biochemistry and medicinal chemistry and a promising component for the preparation of nanocomposite materials. It is a nonessential amino acid component and a precursor of the catecholamine neurotransmitters such as dopamine, norepinephrine, and epinephrine. In opposite to these transmitters, it can cross the blood–brain barrier, which has made it the most effective drug for the symptomatic control of Parkinson's disease. Meso-corticolimbic dopamine (DA) has an important role in the cognitive processes of working memory and reward-based learning.[84,85] Almost half of the patients treated with L-Dopa develop dyskinesia, but the use of dopamine agonists is associated with a lower incidence.[86]

10.2 CONCLUSION

The natural product exists everywhere. When we think about the natural product, the first thing comes into mind is a large plant, but shrubs and bioactive animal products or even bacteria, fungi, yeast, molds that also provide an excellent source of material in modern therapeutics. The products that are naturally derived play an important role in the development of synthetic prodrug molecules for various therapeutic purposes. In this chapter, 15 prodrugs from natural resources were discussed comprehensively. Various synthetic prodrugs have been discovered so far with several added advantages but prodrugs from natural sources have fewer side effects as compared to the synthetic prodrugs along with easy accessibility. However, it will take years to receive attention toward these products to come into mainstream pharmacotherapeutics.

KEYWORDS

- **natural**
- **herbal**
- **prodrugs**
- **microorganism**
- **phytochemical**
- **therapeutics**

REFERENCES

1. Karaman, R. The Prodrug Naming dilemma. *Drug Design* **2013**, *2*(2), 1–3.
2. Kumar, S. V. et al. An Update on Prodrugs From Natural Products. *Asian Pac. J. Med.* **2014**, *7*(1), 54–59.
3. Nikhade, N. et al. A Review of Natural Antioxidant in Medicinal Plant. *Int. J. Pharm. Drug Anal.* **2019**, *7*(2), 11–15.
4. Ansari, M. H. et al. A Short Overview on Anti-Diabetic Natural Products: Reviewing the Herbo-Therapeutics Potential. *Pharm. Sci. Technol.* **2019**, 1–10.
5. J. Padmavathy et al. Natural Product as a Source of Prodrugs. *Bangladesh J. Pharmacol.* **2017**, *12*, 151–161.
6. Ansari, M. H. et al. A Review on Inflammatory Bowel Disease and Their Treatment. *Int. J. Sci. Res.* **2019**, *8*(9), 629–633.
7. Mark Wainwright et al. On the 75th Anniversary of Prontosil. *Dyes Pigments* **2011**, *88*, 231–234.
8. Françoise van Bambeke, Paul M. Tulkens. Infectious Diseases (Fourth Edition), 2017.
9. Bedford, A. et al. Implications of butyrate and its derivatives for gut health and animal production. Animal Nutrition. **2018**, *4*, 151–159.
10. Thao Duy Nguyen et.al. Effects of Monobutyrin and Tributyrin on Liver Lipid Profile, Caecal Microbiota Composition and SCFA in High-Fat Diet-Fed Rats. *J. Nutr. Sci.* **2017**, *6*(51), 1–14.
11. Namkung, H. et al. Antimicrobial Activity of Butyrate Glycerides Toward Salmonella Typhimurium and Clostridium Perfringens. *Poul. Sci. Assoc. Inc.* **2011**, *90*, 2217–2222.
12. Bosmans, J. W. A. M. et al. Comparison of Three Different Application Routes Of Butyrate To Improve Colonic Anastomotic Strength In Rats. *Int. J. Colorectal Dis.* **2016**, 1–9.
13. Tordjman, S. et al. Melatonin: Pharmacology, Functions and Therapeutic Benefits. *Curr. Neuropharm.* **2017**, *15*, 434–443.

14. Tarocco, A. et al. Melatonin as a Master Regulator of Cell Death and Inflammation: Molecular Mechanisms and Clinical Implications for Newborn Care. *Cell Death Dis.* **2019**, *10*(317), 1–12.
15. Xiea, Z. et al. A Review of Sleep Disorders and Melatonin. *Neurol. Res.* **2017**, 1–7.
16. Yang, L. P. H. Romidepsin, In the Treatment of T-Cell Lymphoma. *Drugs.* **2011**, *71*(11), 1469–1480.
17. VanderMolen, K. M. et al. Romidepsin (Istodax, NSC 630176, FR901228, FK228, depsipeptide): a Natural Product Recently Approved for Cutaneous T-Cell Lymphoma. *J. Antibiot.* **2011**, *64*, 525–531.
18. Yang, L. P. H. et al. Romidepsin: A Guide to Its Clinical Use in Cutaneous T-Cell Lymphoma. *Am. J. Clin. Dermatol.* **2012**, *13*(1), 67–71.
19. Kwong, Y.-L. Alemtuzumab Induced Complete Remission of Romidepsin-Refractory Large Cell Transformation of Mycosis Fungoides. *Ann. Hematol.* **2013**, 1–2.
20. Hegarty, S. V. et al. Romidepsin Induces Caspase-Dependent Cell Death in Human Neuroblastoma Cells. *Neurosci. Lett.* 2017, 1–18.
21. Zhang, C. et al. Purification and Characterization of Baicalin-b-D-glucuronidase Hydrolyzing Baicalin to Baicalein from Fresh Roots of Scutellaria Viscidula Bge. *Process Biochem.* **2005**, *40*, 1911–1915.
22. Sun, Z. et al. Electrochemical Investigations of Baicalin and DNA–Baicalin Interactions. *Anal. Bioanal. Chem.* **2004**, *379*, 283–286.
23. Wena, Y.-F. et al. Baicalin Prevents Cadmium Induced Hepatic Cytotoxicity, Oxidative Stress and Histomorphometric Alterations. *Exp. Toxicol. Pathol.* **2013**, *65*, 189–196.
24. Ohkoshi, E. et al. Simple Preparation of Baicalin from Scutellariae Radix. *J. Chromatogr. A.* **2009**, *1216*, 2192–2194.
25. Min, Li-Weber. New Therapeutic Aspects of Flavones: The Anticancer Properties of Scutellaria and Its Main Active Constituents Wogonin, Baicalein and Baicalin. *Cancer Treat. Rev.* **2009**, *35*, 57–68.
26. Hao, H. et al. Baicalin Suppresses Expression of Chlamydia Protease-Like Activity Factor in Hep-2 Cells Infected by Chlamydia Trachomatis. *Fitoterapia* **2009**, *80*, 448–452.
27. Zhou, B.-R. et al. Baicalin Protects Human Fibroblasts Against Ultraviolet B-Induced Cyclobutane Pyrimidine Dimers Formation. *Arch. Dermatol. Res.* **2008**, *300*, 331–334.
28. Horn-Ross, P. L. et al. Assesşing Phytoestrogen Exposure in Epidemiologic Studies: Development of a Database (United States). *Cancer Causes Control* **2000**, *11*, 289–298.
29. AlexanderVSirotkin. Phytoestrogens and Their Effects. *Eur. J. Pharm.* **2014**, 1–7.
30. Dixon, R. A. Phytoestrogens. *Annu. Rev. Plant Biol.* **2004**, *55*, 225–261.
31. Murkies, A. L. et al. Clinical Review (Phytoestrogen). *J. Clin. Endocrinol. Metab.* **1998**, *83*(2), 297–303.
32. Eden, J. A. Phytoestrogens for Menopausal Symptoms: A Review. *Maturitas* **2012**, *72*, 157–159.
33. Ziegler, R. G. Phytoestrogens and Breast Cancer. *Am. Soc. Clin. Nutr.* **2004**, *79*, 183–184.

34. Albertazzi, P. et al. Dietary Soy Supplementation and Phytoestrogen Levels. *Obstet. Gynaecol.* **1999**, *94*(2), 229–231.
35. Raju, S. et al. Effect of Light Intensity on Photosynthesis and Accumulation of Sennosides in Plant Parts of Senna (Cassia angustifolia Vahl.). *Ind. J. Plant Physiol.* **2013**, *18*(3), 285–289.
36. Mukhopadhyay, M. J. et al. Genotoxicity of Sennosides on the Bone Marrow Cells of Mice. *Food Chem. Toxicol.* **1998**, *36*, 937–940.
37. Siddique, I. et al. Stimulation of In Vitro Organogenesis from Epicotyl Explants and Successive Micropropagation Round in Cassi Angustifolia Vahl.: An Important Source of Sennosides. *Agroforest Syst.* **2013**, *87*, 583–590.
38. Sun, S.-W. et al. Validated HPLC Method for Determination of Sennosides A and B in Senna Tablets. *J. Pharm. Biomed. Anal.* **2002**, *29*, 881–894.
39. Choi, S. B. et al. Insulin Sensitizing and a-Glucoamylase Inhibitory Action of Sennosides, Rheins and Rhaponticin in Rhei Rhizoma. *Life Sci.* **2006**, *78*, 934–942.
40. Wang, Y.-R. et al. Barbaloin Loaded Polydopamine-Polylactide-TPGS (PLA-TPGS) Nanoparticles Against Gastric Cancer as a Targeted Drug Delivery System: Studies In Vitro and In Vivo. *Biochem. Biophys. Res.* **2018**, 1–9.
41. Mangaiyarkarasi, S. P. et al. Benefits of Aloe Vera in Dentistry. *J. Pharm. Bioallied Sci.* **2015**, *7*(1), 255–259.
42. Peiyong, Z. et al. Barbaloin Pretreatment Attenuates Myocardial Ischemia-Reperfusion Injury Via Activation of AMPK. *Biochem. Biophys. Res. Commun.* **2017**, *490*, 1215–1220.
43. Choi, S. et al. A Review on the Relationship Between Aloe Vera Components and Their Biologic Effects. *Semin. Integr. Med.* **2003**, *1(1)*, 53–62.
44. Duarte, E. L. et al. On the Interaction of the Anthraquinone Barbaloin with Negatively Charged DMPG Bilayers. *Langmuir* **2008**, *24*, 4041–4049.
45. Teruko, E. et al. In-Vitro and In-Vivo Evaluation of Enhancing Activity of Glycyrrhizin on the Intestinal Absorption of Drug. *Pharm. Res.* **1999**, *16(1)*, 80–86.
46. Nokhodchi, A. et al. The Effect of Glycyrrhizin on the Release Rate and Skin Penetration of Diclofenac Sodium from Topical Formulations. *Farmaco* **2002**, *57*, 883–888.
47. Ashfaq, U. A. et al. Glycyrrhizin as Antiviral Agent Against Hepatitis C Virus. *J. Transl. Med.* **2011**, 109–112.
48. Imai, K. et al. Radical Scavenging Ability of Glycyrrhizin. *Free Radic. Antioxid.* **2013**, *3*, 40–42.
49. Shiota, G. et al. Inhibition of Hepatocellular Carcinoma by Glycyrrhizin in Diethyl-Nitrosoamine Treated Mice. *Carcinogenesis* **1999**, *20*(1), 59–63.
50. Wan, Xu-yinga et al. Hepato-protective and anti-hepatocarcinogenic effects of glycyrrhizin and matrine. *Chem. Biol. Inter.* **2009**, *181*, 15–19.
51. Henwood, J. M. et al. Lovastatin, A Preliminary Review of its Pharmacodynamic Properties and Therapeutic Use in Hyperlipidaemia. *Drugs* **1988**, *36*, 429–454.
52. Gladys, C. et al. Efficacy and Tolerability of Policosanol Compared with Lovastatin in Patients with Type-II Hypercholesterolemia and Concomitant Coronary Risk Factors. *Curr. Ther. Res.* **2000**, *61*(3), 137–145.
53. Padayatty, S. J. et al. Lovastatin-Induced Apoptosis in Prostate Stromal Cells. *J. Clin. Endocrinol. Metab.* **1997**, *82*(5), 1434–1439.

54. Williams, J. P. et al. Effect of Administration of Lovastatin on the Development of Late Pulmonary Effects after Whole-Lung Irradiation in a Murine Model. *Radiat. Res.* **2004**, *161*, 560–567.
55. Chongwoo, Y. et al. Human, Rat, and Mouse Metabolism of Resveratrol. *Pharm. Res.* **2002**, *19*(12), 1907–1914.
56. Milosz, J. et al. Resveratrol in Prostate Diseases – A Short Review. *Cent. Eur. J. Urol.* **2013**, *66*, 144–149.
57. Baur, J. A. et al. Resveratrol Improves Health and Survival of Mice on a High-Calorie Diet. *Nature* **2006**, *444*, 337–341.
58. Emilia Juan, M. et al. Trans-Resveratrol, a Natural Antioxidant from Grapes, Increases Sperm Output in Healthy Rats. *Am. Soc. Nutr. Sci.* **2005**, *135*(4), 757–760.
59. Valerie, V. et al. Therapeutic Potential of Resveratrol in Alzheimer's Disease. *BMC Neurosci.* **2008**, *9*(2), 01–05.
60. Bastianetto, S. et al. Neuroprotective Action of Resveratrol. *Bioch. Biophys. Acta* **2014**, 01–07.
61. James, A. Crowell et al. Resveratrol-Associated Renal Toxicity. *Toxicol. Sci.* **2004**, *82*, 614–619.
62. Dong, Z. Molecular Mechanism of the Chemo-Preventive Effect of Resveratrol. *Mutat. Res.* **2003**, 145–150.
63. Lin, D. et al. Ursolic Acid Induces U937 Cells Differentiation by PI3K/Akt Pathway Activation. *Chin. J. Nat. Med.* **2014**, *12*(1), 0015–0019.
64. Schneider, P. et al. Rapid Solubility Determination of the Triterpenes Oleanolic Acid and Ursolic Acid by UV-Spectroscopy in Different Solvents. *Phytochem. Lett.* **2009**, *2*, 85–87.
65. Yong, S. L. et al. Inhibition of Ultraviolet-A-Modulated Signaling Pathways by Asiatic Acid and Ursolic Acid in HaCaT Human Keratinocytes. *Eur. J. Pharm.* **2003**, *476*, 173–178.
66. Litao, L. et al. Ursolic Acid Promotes the Neuro-Protection by Activating Nrf2 Pathway After Cerebral Ischemia in Mice. *Brain Res.* **2013**, *1497*, 32–39.
67. Jing-song, W.; Ren, T.-N. ; Tao, X. Ursolic Acid Induces Apoptosis by Suppressing the Expression of FoxM1 in MCF-7 Human Breast Cancer Cells. *Med. Oncol.* **2012**, *29*(1), 10–15.
68. Akihisa, T. et al. Cytotoxic Activity of Perilla Frutescens Var. Japonica Leaf Extract is Due to High Concentrations of Oleanolic and Ursolic Acids. *J. Nat. Med.* **2006**, *60*, 331–333.
69. Patricia, k. w. et al. Antioxidant Activity of the Flavonoid Hesperidin in Chemical and Biological Systems. *J. Agric. Food Chem.* **2005**, *53*, 4757–4761.
70. Li, W. et al. Hesperidin, a Plant Flavonoid Accelerated the Cutaneous Wound Healing in Streptozotocin-Induced Diabetic Rats: Role of tgf-β/smads and ang-1/tie-2 Signaling Pathways. *EXCLI J.* **2018**, *17*, 399–419.
71. Tirkey, N. et al. Hesperidin, a Citrus Bioflavonoid, Decreases the Oxidative Stress Produced by Carbon Tetrachloride in Rat Liver and Kidney. *BMC Pharmacol.* **2005**, *5*(2), 01–08.
72. Jung, U. J. et al. The Hypoglycemic Effects of Hesperidin and Naringin Are Partly Mediated by Hepatic Glucose-Regulating Enzymes in C57BL/KsJ-db/db Mice. *Am. Soc. Nutr. Sci.* **2004**, *134*(10), 2499–2503.

73. zhao, J. et al. Hesperidin Inhibits Ovarian Cancer Cell Viability Through Endoplasmic Reticulum Stress Signaling Pathways. *Oncol. Lett*. **2017**, *14*, 5569–5574.
74. chen, M.-C. et al. Hesperidin Upregulates Heme Oxygenase-1 To Attenuate Hydrogen Peroxide-Induced Cell Damage in Hepatic L02 Cells. *J. Agric. Food Chem.* **2010**, *58*, 3330–3335.
75. Ho, S.-C. et al. Hesperidin, Nobiletin, and Tangeretin are Collectively Responsible for the Anti-Neuroinflammatory Capacity of Tangerine Peel (Citri reticulatae pericarpium). *Food Chem. Toxicol.* **2014**, *71*, 176–182.
76. Roohbakhsh, A. et al. Neuropharmacological Properties and Pharmacokinetics of the Citrus Flavonoids Hesperidin and Hesperetin-A Mini-Review. *Life Sci.* **2014**, *113*, 1–6.
77. Jun, S.-Y. et al. Inhibitory Effects of Arbutin-b-glycosides Synthesized from Enzymatic Transglycosylation for Melanogenesis. *Biotechnol. Lett.* **2008**, *30*, 743–748.
78. Migas, P. et al. The Significance of Arbutin and Its Derivatives in Therapy and Cosmetics. *Phytochem. Lett.* **2015**, *962*, 01–06.
79. Tokiwa, Y. et al. Enzymatic Synthesis of Arbutin Undecylenic Acid Ester and Its Inhibitory Effect on Mushroom Tyrosinase. *Biotechnol. Lett.* **2007**, *29*, 481–486.
80. Tokiwa, Y.; Kitagawa, M., Raku, T. Enzymatic Synthesis of Arbutin Undecylenic Acid Ester and Its Inhibitory Effect on Mushroom Tyrosinase. *Biotechnol. Lett.* **2007**, 29 (3), 481–486.
81. Yea, J. et al. Arbutin Attenuates LPS-Induced Lung Injury Via Sirt1/Nrf2/NF-κBp65 Pathway. *Pulm. Pharmacol. Ther.* **2019**, *54*, 53–59.
82. Piekoszewska, A. et al. Arbutin production in Ruta graveolens L. and Hypericum perforatum L. In Vitro Cultures. *Acta Physiol. Plant.* **2010**, *32*, 223–229.
83. Lee, H.-J. et al. Anti-inflammatory effects of arbutin in lipopolysaccharide-stimulated BV2 microglial cells. *Inflamm. Res.* **2012**, *61*, 817–825.
84. Jaber, M.; Lambert, J.-F.-O. A New Nanocomposite: L-DOPA/ Laponite. *J. Phys. Chem. Lett.* **2010**, *1*, 85–88.
85. Cools, R. et al. L-DOPA Disrupts Activity in the Nucleus Accumbens during Reversal Learning in Parkinson's Disease. *Neuropsychopharmacology* **2007**, *32*, 180–189.
86. Jenner, P. Molecular Mechanisms of L-DOPA-Induced Dyskinesia. *Nat. Rev. Neurosci.* **2008**, *9*, 665–677.

CHAPTER 11

Perspectives of Nature-Oriented Pharmacotherapeutics for the Effectual Management of Hemorrhoidal Symptoms

TARANPREET KAUR BAMRAH[1], MOJABIR HUSSEN ANSARI[2], and DEBARSHI KAR MAHAPATRA[3*]

[1]*Department of Pharmaceutics, Gurunanak College of Pharmacy and Technical Institute, Nagpur 440026, India*

[2]*Department of Quality Assurance, Gurunanak College of Pharmacy and Technical Institute, Nagpur 440026, India*

[3]*Department of Pharmaceutical Chemistry, Dadasaheb Balpande College of Pharmacy, Nagpur 440037, India*

[*]*Corresponding author. E-mail: mahapatradebarshi@gmail.com*

ABSTRACT

Hemorrhoids can be said to be as one of the most prevalent ailments in modern days which have been linked with noteworthy impact on the quality of life. Management of this disease by synthetic drugs and modern-day surgeries has not shown any impressive results and has limited therapeutic options. The herbal products are Mother Nature's gift. This chapter comprehensively discusses the management of hemorrhoid disease and its associated symptoms using various natural pharmacotherapeutic options such as *Aesculus hippocastanum*, *Allium cepa*, *Bergenia ciliate*, *Bergenia ligulata*, *Bergenia stracheyi*, *Boswellia serrata*, *Boswellia carterii*, *Brassica rapa*, *Cestrum auriculatum*, *Cestrum hediundinum*, *Cissus quadrangularis*, *Commiphora mukul*, *Commiphora myrrha*, *Euphorbia prostrate*, *Ficus carica*, *Ginkgo biloba*, *Hamamelis virginiana*, *Juniperus oxycedrus*, *Juniperus sabina*, *Juniperus polycarpos*, *Juniperus communis*, *Mangifera indica*, *Melastoma malabathricum*, *Momordica charantia*, *Myrtus communis*, *Onosma* species, *Oryza sativa*, *Phlomis* species, *Plantago ovata*, *Raphanus sativus*, *Ruscus aculeatus*, *Sesamum indicum*, *Syzygium cumini*, *Terminalia chebula*, *Triticum aestivum*, *Vaccinium myrtillus*, *Verbascum mucronatum*, *Verbascum latisepalum*, *Verbascum salviifolium*, *Verbascum lasianthum*, *Verbascum pterocalycinum*, and *Zingiber officinale*.

11.1 INTRODUCTION

Hemorrhoid is a very common disorder. Hemorrhoids are not veins in true meaning, but these are the vascular spaces present in the anal canal from the time of birth itself. These spaces form cushion above the anal canal and help in the control of continence and the passing of the gas. They become problematic only when engorged and may prolapse which results in bleeding, pain, or itching. This takes place when the connective and the smooth muscle of the venous sinuses, stretch and break.[1]

When the process of defecation is carried out, intra-abdominal pressure is opposed by the atmospheric pressure which leads to the development of a shearing force that is accentuated by prolonged or excessive straining. This can be due to faulty diets, particularly low in fiber contents. These hemorrhoidal plexi experience this shearing force leading to the destruction of the supportive tissue. During straining, there is a transudation of the blood which causes bleeding (referred to as Stage-I). Continual stretching leads to the prolapse of the submucosa which in the beginning returns spontaneously (referred to as Stage-II). With the increasing disruption, the manual reduction is needed (referred to as Stage-III) and finally, the mucosal suspensory ligament attaches mucosa to the muscular wall (referred to as Stage-IV). There is free communication between these two plexi and a hemorrhoid is formed which is irreducible (Table 11.1).[2]

TABLE 11.1 Internal Hemorrhoids their Classification and Symptoms.

Stages	Prolapse	Symptoms
I	None	Bleeding
II	Prolape happens with movement	Prolapse, bleeding, mild discomfort
III	Prolapse happens with bowel movements	Prolapse, bleeding, occasional pruritis, discharge, discomfort
IV	Persistent, reduction not possible	Prolapse, bleeding, pain, thrombosis, soiling, discharge

Hemorrhoids may be internal or external based on their location of occurrence. The external hemorrhoids can be easily seen and felt around the anus. There are various factors which can be a cause of hemorrhoids. The most common factors include low fiber diets which lead to constipation and increased straining during stool passing. Pregnancy can also be the cause; however, the etiology is unknown. Prolonged sitting is also found to be a reason. The external hemorrhoids are usually formed due to the blood clot around the anus.[3]

The diagnosis of the external hemorrhoids is easy as it can be seen from the outside itself. Bleeding is the common symptom of hemorrhoids; however, it can also be due to colorectal cancer. Swollen rectum and anus blood vessels are first checked by the medical practitioners as the indexed diagnosis process. A digital rectal examination is done using a gloved lubricated finger. Anoscope and proctoscope are also used in the rectal examination. Sigmoidoscope and colonoscope are concurrently used for the examination of the colon.[4]

11.2 THERAPEUTIC ARMAMENTARIUM FROM MOTHER NATURE

Nature has provided us with the treasure trove of remedies to treat different ailments in the form of medicinal plants. Nearly 6000 plants are present with medicinal importance in the Indian subcontinent, representing 73% of the total medicinal needs of the world. Botanical treatments and nutritional therapies are quite safe, offer low side effects, and are effective for the treatment of hemorrhoids. Several botanical extracts have been shown to improve capillary flow, microcirculation, vascular tone, and strengthen the connective tissue of the perivascular substrate in hemorrhoids. In addition to it, the presence of pharmacological properties like antioxidant, anti-inflammatory, antiedema, and hepatoprotective provides additional relief in the therapeutic regimen of hemorrhoids.[5] Following secondary metabolites of plants are reported as the responsible factor for expressing the antihemorrhoidal properties[6,7]:

- **Triterpenoids:** *Centella asiatica* contains active constituents like pentacyclic derivatives such as asiatic acid, madecassic acids, and asiaticoside which helps in hemorrhoid treatment by improving connective tissue integrity, elevate antioxidant levels in wound healing, improvement in capillary permeability in CVI, and varicose veins treatments.
- **Tannins:** Plants such as *Hamamelis virginiana*, *Emblica officinalis*, *Achillea milletolium*, *Terminalia chebula*, and *Vitex nigundo* are rich in tannins which produce anti-inflammatory, astringent, and vasoconstrictive effects that are beneficial in the hemorrhoidal treatment.
- **Saponin glycosides:** The saponin glycosides (which show anti-inflammatory and astringent properties) present in *Ruscus aculeatus* is most commonly used for the treatment of internal hemorrhoids.
- **Triterpenic saponins:** *Aesculus hippocastanum* is a common example containing triterpenic saponins which can be useful for treating hemorrhoids. The in vitro studies have shown that triterpenic saponins inhibit the activities of the enzymes elastase and hyaluronidase which are involved in proteoglycan degradation, thereby compromising the part of the capillary endothelium and extravascular matrix. These properties make it very ideal for hemorrhoids treatment. It also shows venotonic, vasculoprotective, anti-inflammatory, and free radical scavenging properties which additionally play a role in managing bleeding hemorrhoids.
- **Bioflavonoids:** It exhibits phlebotonic activity, vasculoprotective effects, and antagonism of biochemical mediators of inflammation which has additional effects on hemorrhoids. Studies have also shown that it improves venous tone and vein elasticity.
- **Tocotrienols:** They are commonly present in palm oil and rice bran oil which has an indirect effect on hemorrhoids due to antioxidant activity, anticancer effect, and protection to blood vessels.

- **Volatile oils:** The terpenes and their oxygenated compounds are found to be quite useful in the treatment of hemorrhoids.

Some homemade natural remedies that can be used as a part of the pharmacological treatment include[8]:

- In yogurt, black powdered mustard is added and taken along with buttermilk.
- Peels of pomegranate are soaked in the bowl of water and then boiled to get a concoction which is further stored in the bottle and administered once in the evening and morning.
- Horse chestnut extract helps reduce the hemorrhoids as it has anti-inflammatory, anti-edema, and venotonic actions.
- Eating fiber-rich diets is seldom recommended.
- The juice of white radish is extracted, mixed with honey, and applied on the affected area.
- Mashed banana can be taken along a glass of milk to help treat the pain during the passing of the stools.
- Three to four figs are soaked in a glass of water overnight and then taken early in the morning, in an empty stomach.
- Consume more fruits and vegetables such as dry figs, Indian gooseberry, papaya, and radish.
- Yoga poses such as cobra pose and bow pose also help relieve hemorrhoids.
- Indian style toilets should be preferred because the squatting position can help for natural defecation.
- Drinking plenty of water helps to flush out toxins and waste substances from the body.

11.3 HERBAL BASED MANAGEMENT OF HEMORRHOIDAL SYMPTOMS[9-31]

11.3.1 AESCULUS HIPPOCASTANUM (1)

The extracts of the seeds of *Aesculus hippocastanum* are used mainly in Europe to treat hemorrhoidal disease (both internally and externally). Several clinical trials have been conducted on this subject over the years which highlighted that the clinical symptoms were successfully managed after 1 week of administration to the patients under trial along with improved physical state of hemorrhoids. The therapeutically active constituent present in the seed extracts is aescin, a combination of triterpene saponins which exists in two forms α and β, of which, β-aescin is the most active one. Aescin also demonstrates anti-inflammatory, venotonic, and antiedematous activities which have indirect applications in treating the symptoms of hemorrhoids (Fig. 11.1).

11.3.2 ALLIUM CEPA (2)

Allium cepa (onion), a medicine of importance, as stated in traditional Persian medicine, is essentially employed for treating both dry piles and bleeding hemorrhoids due to its analgesic and anti-inflammatory activities. Traditionally, 30 g of onion is finely rubbed with water and 60 g of sugar is added to it to form a formulation. It is then administered twice daily to the hemorrhoid patients.

11.3.3 BERGENIA SPECIES (3)

The rhizometic herb *Bergenia* species (*Bergenia ciliate*, *Bergenia ligulata*, and *Bergenia stracheyi*) is used in the form of root tinctures and rhizome leaf powder for treating hemorrhoids. It contains active phytoconstituents such as tannic acid, gallic acid, glucose, mucilage, wax, metarbin, albumen, mineral salts, bergenin, (+)-catechin, and gallicin. This plant is found in Afghanistan, India, Pakistan, and Himalayan plains and is ethnically employed by the local tribes for treating hemorrhoids in the form of powder and tincture.

11.3.4 BOSWELLIA SERRATA (4)

The gum resins of *Boswellia serrata* Roxb and other species *B. carterii* Bird have been used for years in the traditional Iranian medicine for the treatment of hemorrhoids. Folk medicines, for centuries, have indicated toward gum resin extract of this plant which is extremely useful in treating the chronic inflammatory diseases. In the modern-day, it is found that boswellic acids are pentacyclic triterpenes present in *Boswellia* that shows both potent anti-inflammatory and anticancer activities.

11.3.5 BRASSICA RAPA (5)

The leaves of *Brassica rapa* (turnip) are used for the treatment of hemorrhoids and are a part of ancient traditional Persian medicine literature. They are commonly grown, widely adapted root crops, as a general farm crop, truck crop, or a home garden crop. The juice of the turnip leaves is extracted and mixed with equal quantities of juices of watercress, spinach, and carrots, and then given to the patients for better results. It is also required to take a proper diet of raw fruits and raw vegetables along with the administration of this juice.

11.3.6 CESTRUM AURICULATUM (6)

Peruvian traditional medicine "Hierba santa" uses the *Cestrum* species; *Cestrum auriculatum* and *C. hediundinum* in the treatment of hemorrhoids. These two species have additional anti-inflammatory, astringent, and analgesic effects which made them extremely popular in the ethnomedicinal use because it is essential and additionally needed by the patients suffering from various stages of hemorrhoids.

11.3.7 CISSUS QUADRANGULARIS (7)

Cissus quadrangularis is commonly found in Asia and Africa and is useful particularly in the relief of bleeding hemorrhoids. The plant is extremely useful in treating pain (by selectively inhibiting the local mediators and nociceptors in the central nervous system) and inflammation (reduction in the release and synthesis of related mediators, principally prostaglandins). On longer application, it is perceived that the magnitude of hemorrhoids condenses and prevents bleeding.

11.3.8 COMMIPHORA MUKUL (8)

The gum resins of *Commiphora mukul* and *C. myrrha* are beneficial in the treatment of hemorrhoids and are reported as a plant of importance in traditional Iranian medicine. Terpenoids and guggulusteroids are the active constituents present in this resin which imparts anti-inflammatory and antinociceptive activities.

11.3.9 EUPHORBIA PROSTRATA (9)

Euphorbia prostrate is used for the management of grade-I and grade-II symptomatic hemorrhoids. The exact mechanism of action of this plant in treating hemorrhoids includes an increase in the lymphatic drainage, reduction in the capillary permeability, improvement of the venous tone, protection of capillary bed microcirculation, and inhibition of certain inflammatory reactions. The pharmacological activity is mediated by the presence of flavonoids, phenolic compounds, tannins, etc.

11.3.10 FICUS CARICA (10)

Ficus carica is very effective in relieving constipation and is found to be very helpful in treating hemorrhoids. It also shows potent anti-inflammatory activity which has additional importance in managing the bleeding hemorrhoids. Traditionally, three or four figs are soaked overnight in pure water after being cleaned thoroughly in hot water. They are administered early in the morning in a fresh stomach along with the water in which they were soaked.

11.3.11 GINKGO BILOBA (11)

Ginkgo biloba leaf extract is very useful in the management of acute hemorrhoidal attacks when used in the combination with troxerutin (a flavonoid) and heptaminol (a vasodilator). The anti-inflammatory and venoprotective activity of *Ginkgo biloba* is highly beneficial in treating hemorrhoids.

11.3.12 HAMAMELIS VIRGINIANA (12)

Hamamelis virginiana (Witch-Hazel) gel is used in curing hemorrhoids or bleeding piles entirely. Applying a little Witch hazel astringent gel around the rectum can be beneficial in reducing the inflammation, curing the pain, preventing the burning sensation, swelling, and discomfort as well as reducing hemorrhage.

11.3.13 JUNIPERUS OXYCEDRUS (13)

Juniperus species (*J. oxycedrus*, *J. sabina*, *J. polycarpos*, and *J. communis*) have their immense significance in Turkish folk medicine and traditional Uygur medicine for the treatment of hemorrhoids. The methanolic extracts of fruit and leaves of *Juniperus* species demonstrated anti-inflammatory and antinociceptive activities due to the presence of active diterpenoids such as hinokiol, etc. which additionally help in the therapeutic practice.

11.3.14 MANGIFERA INDICA (14)

The genus Mangifera (Mango) primarily originated in tropical Asia with the greatest number of species found in Indonesia and the Malay Peninsula. The most cultivated *Mangifera* species, *Mangifera indica* has originated from India. In the traditional medicine of Persia, it is mentioned that the mango seeds are an effective remedy for treating bleeding hemorrhoids. In India, the seeds are collected during the mango season, dried in the shade, powdered, and kept stored for use as antihemorrhoid medicine by administering 1–2 g with or without honey, twice daily.

11.3.15 MELASTOMA MALABATHRICUM (15)

The powdered leaves and roots of *Melastoma malabathricum* have been reported to effectually relieve discomfort in hemorrhoids and also for treating bleeding hemorrhoids. The antinociceptive, anti-inflammatory, wound healing, antidiarrheal, cytotoxic, and antioxidant activities of the plant have additive advantages in the treatment of hemorrhoids. The activities are mediated by the presence of flavonoids and tannins.

11.3.16 *MOMORDICA CHARANTIA (16)*

Momordica charantia (bitter melon or bitter gourd) is a flowering vine belonging to the family Cucurbitaceae. The juice of fresh leaves of bitter melon is very useful in treating bleeding hemorrhoids. Three teaspoons of this leaf juice along with a glass of buttermilk is to be taken every morning for treating hemorrhoidal conditions. The paste of the roots of the bitter gourd plant can also be applied over the hemorrhoids for beneficial results.

11.3.17 *MYRTUS COMMUNIS (17)*

Myrtus communis is a part of traditional Iranian medicine used for treating hemorrhoids. The essential oil present in *M. communis* improves the bleeding, reduces permanent pain, diminishes pain during defecation, decreases anal irritation, lessens anal itching, and anal heaviness in hemorrhoidal type-I and type-II patients. The anti-inflammatory and antinociceptive activities of *M. communis* are mediated due to the presence of flavonoids.

11.3.18 *ONOSMA SPECIES (18)*

In Turkish folk medicine, the usefulness of *Onosma* species for treating hemorrhoids is depicted. The plant contains alkannin, shikonin, flavonoids, ferulic acid, and vanillic acid which are responsible for the mediation of anti-inflammatory (inhibition of lipoxygenase enzyme), wound healing, and analgesic activities.

11.3.19 *ORYZA SATIVA (19)*

Oryza sativa commonly known as rice has very low fiber content and is therefore extremely soothing to the digestive system. A thick gruel of rice, mixed with a glass of buttermilk and ripe banana is given twice a day to the patients for treating hemorrhoidal symptoms.

11.3.20 PHLOMIS SPECIES (20)

In Spanish folk medicine, the usefulness of *Phlomis* species for treating hemorrhoids is illustrated. The phytochemicals such as forsythoside B, alyssonoside (phenylpropanoid compounds), phlomiside (steroid), flavonoids (in trace amounts), etc. play a key role in mediating the antinociceptive and anti-inflammatory effects which have importance in treating hemorrhoids.

11.3.21 PLANTAGO OVATA (21)

Plantago ovata is very beneficial for the patients suffering from hemorrhoids by reducing the bleeding, improving the symptoms, and reducing the hemorrhoidal cushions. This bulking laxative is a stool softener that has therapeutic use for posthemorrhoidectomy complications. *P. ovata* contains a large amount of short-chain fatty acids as well as soluble nonstarch polysaccharides. These polysaccharides, under anaerobic fermentation, produce propionate, butyrate, and acetate components in the intestines which play a critical role in reducing the inflammation of the anorectal region.

11.3.22 RAPHANUS SATIVUS (22)

White radish (*Raphanus sativus*) is known to have ethnopharmacological significance for centuries in the treatment of hemorrhoids and associated bleeding. It is taken systematically by first mixing 100 mg of the content with a teaspoon of honey with a little common salt, in a twice-daily manner. *R. sativus* ground into a paste in milk can also be applied over inflamed pile masses to relieve pain and swelling.

11.3.23 RUSCUS ACULEATUS (23)

Ruscus aculeatus (Butcher's broom) is a member of the Liliaceae family and is a native to Mediterranean Europe and Africa. Butcher's broom, in small quantity, proved to be one of the commonly used herbs in the treatment of hemorrhoids. It is given in the form of a capsule along with trimethylhesperidin chalcone (a flavonoid complex) and ascorbic acid (vitamin C). The steroids present in this plant such as ruscogenins play an effective role in reducing the tenderness, strengthening the local blood vessels as well as preventing excessive bleeding due to anorectal hemorrhoids.

11.3.24 *SESAMUM INDICUM* (24)

The seeds of *Sesamum indicum* L. (Pedaliaceae) are extremely useful in treating hemorrhoids, constipation, pain, and wound healing. They can be taken in the form of a decoction by boiling 20 g of seeds in water till it is reduced by one-third of the quantity, or as sweetmeats. They can also be given with butter, ground to paste with water, and applied over bleeding hemorrhoids.

11.3.25 *SYZYGIUM CUMINI* (25)

The genus *Syzygium* is one of the genera of the myrtle belonging to the family Myrtaceae. It is native to the tropics, particularly to tropical America and Australia. The fruit pulp is an effective remedy for treating bleeding hemorrhoids, when administered with common salt for 2–4 months duration every morning.

11.3.26 *TERMINALIA CHEBULA* (26)

In traditional Tibetan medicine and Ayurvedic materia medica, *Terminalia chebula* is referred to as the "King of Medicines" because of its extraordinary powers of healing. The herb Chebulic myroblan is a popular remedy for piles due to its anti-inflammatory and wound healing properties. *Terminalia chebula* is roasted to the brown color in castor oil, then powdered, and stored for its use in hemorrhoids (half a teaspoon taken at the bedtime). A decoction of the herb is usually prepared by boiling six to seven dry fruits and is employed for washing the bleeding hemorrhoids. The paste of the fruit is mixed with oil and applied for external applications to the anorectal region.

11.3.27 TRITICUM AESTIVUM (27)

Wheat grass (*Triticum aestivum*) juice is used to detoxify the walls of the hemorrhoids (anorectal region) by administrating through an enema. The juice (90–120 mL) is given with lukewarm or neem water through enema for 20 min durations and further retained for 15 min.

11.3.28 VACCINIUM MYRTILLUS (28)

Vaccinium myrtillus (bilberry), also known as European blueberry, huckleberry, whortleberry, or blueberry, is a member of the Ericaceae family. Ethnobotanical and scientific evidence have suggested the applications of bilberry in treating hemorrhoidal symptoms. The presence of bioflavonoid makes this herb useful in curing bleeding piles by improving the blood flow very efficiently.

11.3.29 VERBASCUM SPECIES (29)

Various species of *Verbascum* (*V. mucronatum*, *V. latisepalum*, *V. salviifolium*, *V. lasianthum*, and *V. pterocalycinum*) have their ethnotraditional significance in Turkish medicine for managing type-II hemorrhoids. These Mullein leaf extracts are native to Europe and cultivated in temperate areas of the world, including Hawaii, North America, New Zealand, and Australia. The aqueous extract of *V. mucronatum* flower demonstrates potent anti-inflammatory, anti-nociceptive, and wound healing activities. Fractionation of this herbal tonics leads to isolation of different iridoid glycosides of which verbascoside, aucubin, and ilwensisaponin A are responsible for the pharmacological activities by reducing the dilation of veins.

11.3.30 ZINGIBER OFFICINALE (30)

Zingiber officinale or common cooking ginger has a pivotal role in treating various stages of hemorrhoids. This resin grows in tropical Asia, but is now grown as a commercial crop in Latin America and Africa as well as South East Asia. Half a teaspoon of fresh ginger juice, mixed with one teaspoon of fresh lime juice and fresh mint juice, along with a tablespoon of honey constitutes an effective medicine for treating hemorrhoids.

FIGURE 11.1 Some exclusive plants for the treatment of hemorrhoids.

11.4 CONCLUSION

The chapter highlighted the unknown roles of *Aesculus hippocastanum*, *Allium cepa*, *Bergenia ciliate*, *Bergenia ligulata*, *Bergenia stracheyi*, *Boswellia serrata*, *Boswellia carterii*, *Brassica rapa*, *Cestrum auriculatum*, *Cestrum hediundinum*, *Cissus quadrangularis*, *Commiphora mukul*, *Commiphora myrrha*, *Euphorbia prostrate*, *Ficus carica*, *Ginkgo biloba*, *Hamamelis virginiana*, *Juniperus oxycedrus*, *Juniperus sabina*, *Juniperus polycarpos*, *Juniperus communis*, *Mangifera indica*, *Melastoma malabathricum*, *Momordica charantia*, *Myrtus communis*, *Onosma* species, *Oryza sativa*, *Phlomis* species, *Plantago ovata*, *Raphanus sativus*, *Ruscus aculeatus*, *Sesamum indicum*, *Syzygium cumini*, *Terminalia chebula*, *Triticum aestivum*, *Vaccinium myrtillus*, *Verbascum mucronatum*, *Verbascum latisepalum*, *Verbascum salviifolium*, *Verbascum lasianthum*, *Verbascum pterocalycinum*, and *Zingiber officinale* in the effective management of hemorrhoids. However, in the modern line of therapy, these extracts are not very prevalent, but, with the due course of time, it is expected that these plants will be utilized in the form of medicinal formulations for providing relief from bleeding hemorrhoids.

KEYWORDS

- **hemorrhoids**
- **piles**
- **treatment**
- **management**
- **therapeutics**
- **natural**
- **herbal**
- **mechanism**

REFERENCES

1. Marcon, N. E. Hemorrhoids. *Can. J. Gastroenterol.* **1990,** *4* (9), 554–555.

2. Ahad, H. A. et al. Herbal Treatment for Hemorrhoids. *JITPS* 2010, *1* (5), 236–242.
3. Bharat, G. Botanicals an Alternative Treatment Approach for Hemorrhoids – A Review. *Indian J. Pharmaceutics* **2014,** *5* (1), 37.
4. Abdollahi, M. et al. Evidence-based Review of Medicinal Plants Used for the Treatment of Hemorrhoids. *Int. J. Pharmacol.* **2013,** *9* (1), 1–11.
5. Margină, D. et al. Assessment of the Potential Health Benefits of Certain Total Extracts from *Vitis vinifera, Aesculus hyppocastanum* and *Curcuma longa. Exp. Ther. Med.* **2015,** *10* (5), 1681–1688.
6. Siddiqui, M. Z. *Boswellia Serrata,* A Potential Antiinflammatory Agent: An Overview. *Indian J. Pharm. Sci.* **2011,** *73* (3), 255–261.
7. Kawano, M.; et al. Anti-inflammatory and Analgesic Components from "Hierba Santa," A Traditional Medicine in Peru. *J. Nat. Med.* **2009,** *63* (2), 147–158.
8. Hammond, G.B. et al. A Survey of Traditional Medicinal Plants from the Callejo′n deHuaylas, Department of Ancash, Peru. *J. Ethnopharmacol.* 1996, *61*, 17–30.
9. Panpimanmas, S. et al. Experimental Comparative Study of the Efficacy and Side Effects of Cissus quadrangularis L. (Vitaceae) to Daflon (Servier) and Placebo in the Treatment of Acute Hemorrhoids. *J. Med. Assoc. Thai.* **2010,** *93* (12), 1360.
10. Yousefi, M. et al. Clinical Evaluation of *Commiphora Mukul,* A Botanical resin, in the Management of Hemorrhoids: A Randomized Controlled Trial. *Pharmacogn. Mag.* **2013,** *9* (36), 350–356.
11. Porwal, A. et al. Euphorbia Prostrata - A Clinically Proven Drug in Hemorrhoids " Multiple Pharmacological Actions Targeting Pathological Processes. *Int. J. Med. Health Sci.* **2015,** *4* (2), 269–273.
12. Sumboonnanonda, K. et al. Clinical study of the Ginko biloba--Troxerutin-Heptaminol Hce in the treatment of acute hemorrhoidal attacks. *J. Med. Assoc. Thai.* *87* (2), **2004,** 137–142.
13. Zhao, J. et al. Evaluation on Analgesic and Anti-Inflammatory Activities of Total Flavonoids from Juniperus sabina. *Evid.-Based Complement. Alternat. Med.* **2018,** 9.
14. Joffry, S. M. et al. *Melastoma malabathricum* (L.) Smith Ethnomedicinal Uses, Chemical Constituents, and Pharmacological Properties: A Review. *Evid.-Based Complement. Alternat. Med.* **2011,** *2012.*
15. Mahboubi, M. Effectiveness of *Myrtus communis* in the Treatment of hemorrhoids. *J. Integr. Med.* 2017, *15* (5), 351–358.
16. Kumar, N. et al. Onosma L.: A Review of Phytochemistry and Ethnopharmacology. *Eur. PMC* **2013,** *7* (14), 140–151.
17. Ullah, R. et al. Phlomiside: A New Steroid from *Phlomis bracteosa. Arabian J. Chem.* **2017,** *10* (1), S1303–S130.
18. Mahmood, A. Plantago ovata: A Comprehensive Review on Cultivation, Biochemical, Pharmaceutical and Pharmacological Aspects. *Acta Poloniae Pharm.* **2018,** *74* (3), 739–746.
19. Tatli, I. I. Antinociceptive and Anti-inflammatory Activities of Seven Endemic *Verbascum* Species Growing in Turkey. *J. Pharm. Biol.* **2009,** *46*, 781–788.
20. Hashempur, M. H. et al. An evidence-based Study on Medicinal Plants for Hemorrhoids in Medieval Persia. *J. Evid. Based Complement. Alternat. Med.* **2017,** *22* (4), 969–981.

21. Ahmad, M. et al. *Bergenia ciliata*: A Comprehensive Review of its Traditional Uses, Phytochemistry, Pharmacology and Safety. *Biomed. Pharmacother.* **2018,** *97*, 708–721.
22. Hashempur, M. H. et al. An Evidence-Based Study on Medicinal Plants for Hemorrhoids in Medieval Persia. *J. Evid. Based Integr. Med.* 2017, *22* (4), 969–981.
23. Badgujar, S. B. et al. Traditional Uses, Phytochemistry and Pharmacology of *Ficus carica*: A Review. *J. Pharm. Biol.* 2014, *52* (11), 1487–1503.
24. Thring, T. S. A. et al. Antioxidant and Potential Anti-inflammatory Activity of Extracts and Formulations of White Tea, Rose, and Witch Hazel on Primary Human Dermal Fibroblast Cells. *J. Inflamm.* **2011,** *8*.
25. Ahmad, N. et al. Momordica Charantia: For Traditional Uses and Pharmacological Actions. *J. Drug Delivery Ther.* **2016,** *6* (2), 1–3.
26. Abascal, K.; Yarnell, E. Botanical Treatments for Hemorrhoids. *Alternat. Complement. Therap.* **2005,** *11* (6), 285–289.
27. Saleem, T. S. M. et al. Phyto-pharmacological Review of Sesamum indicum Linn. *Nat. Prod.* **2009,** *5* (4), 184–190.
28. Jagetia, G. C. Phytochemical Composition and Pleotropic Pharmacological Properties of Jamun, Syzygium Cumini Skeels. *J. Explor. Res. Pharmacol.* 2017, *2*, 54–66.
29. Bag, A. The Development of *Terminalia chebula* Retz. (Combretaceae) in Clinical Research. *Asian Pac. J. Trop. Biomed.* **2013,** *3* (3), 244–252.
30. Rajani, C. et al. Golden Herbs used in Piles Treatment: A Concise Report. *Int. J. Drug Dev. Res.* **2012,** *4* (4), 50–68.
31. Sun, Z. et al. Review of Hemorrhoid Disease: Presentation and Management. *Clin. Colon Rectal Surg.* **2016,** *29* (1). 22–29.

CHAPTER 12

Effect of Light on Transport of Potassium Thiocyanate in Aqueous Solutions

SÓNIA I. G. FANGAIA[1], PEDRO M. G. NICOLAU[1], FERNANDO A. D. R. A. GUERRA[1], V. M. M. LOBO[2], and ANA C. F. RIBEIRO[2*]

[1]*Faculty of Medicine, University of Coimbra, Av. Bissaya Barreto, Blocos de Celas, Coimbra 3000, Portugal*

[2]*Department of Chemistry, University of Coimbra, Coimbra 3004-535, Portugal*

**Corresponding author. E-mail: anacfrib@ci.uc.pt*

ABSTRACT

By using a conductometric technique, diffusion coefficients have been measured for potassium thiocyanate in aqueous solutions of different concentrations at 25 °C in a dark chamber. These data are compared with the data obtained in the presence of sunlight using the same technique and with the data obtained from another research group.

These data are of enormous importance in many applied areas, such as corrosion studies in the oral cavity of humans, helping to better understand the complex structure of these systems.

12.1 INTRODUCTION

Data of transport properties of electrolytes in different media is very useful in both fundamental[1–10] and applied areas, such as biomedical and pharmaceutical applications,[4–8] and corrosion studies.[11–18]

Our research group is especially focused on dental restoration research; hence, in getting data for ionic systems implicated in the dental damage processes into the oral cavity. These data are not available in the literature but are essential to adequately understand and resolve some of the problems originating from the deterioration of the dental restorations. Among the electrolytes in human saliva, potassium thiocyanate (KSCN) has been widely studied because of its diverse properties, such as antioxidant ones, protecting cells against oxidizing agents, for example, hypochlorous acid (HOCl), making it a potentially useful therapeutic agent.[19]

Thus, due to thiocyanate's (SCN's) diverse properties, it has driven us to characterize the transport of these salts in solutions with and without light.

However, to our knowledge, research of the transport behavior of this electrolyte under different nonstatic storage environments is still limited in the literature. In fact, few have taken into account these studies in the oral cavity.[4,8]

Among the transport properties of the solutions (diffusion, viscosity, transport number, and conductance), we focused mainly on the study of the interdiffusion coefficients in aqueous electrolyte solutions, a subject on which we have been developing experimental work.[20–46] In fact, having in mind the study of the behavior of the mutual diffusion of ionic systems possibly resulting from the corrosion of metals used in dentistry, oral cavity conditions have been simulated with special attention to the effects of some constituents of foods and beverages, as well as some components of oral hygiene products under different conditions: concentrations, temperatures, pH, techniques used, and different media (water, water with different saliva constituents, e.g., lactic acid,[40] copper (II) chloride, and some carbohydrates: sucrose, glucose, fructose, and macrocycles[45]). The quantitative measure of this property can be obtained through the diffusion coefficient of each component in solution. If there is a concentration gradient in a binary electrolyte solution, then the constraint of maintaining electrical neutrality ensures that positive and negative ions move from the region of higher to lower concentration at the same speed. On a volume-fixed and most other reference frames, there is only one mutual or interdiffusion coefficient in this system.

Our group was dedicated in the last years to the study of the diffusion behavior of binary systems, that is, chemical systems containing electrolytes (e.g., ions resulting from corrosion and wear of dental material, such as chromium and nickel), in different media (e.g. water, saliva, decomposition of food and changes in pH) (see Table 1 from Ref. [8]).

Having in mind that data on differential mutual diffusion coefficients, *D*, for KSCN in absence of light have not been published; in this part of the present work we have presented and discussed our measured diffusion coefficients of KSCN in solutions at finite concentrations (i.e., from 0.010 to 0.200 mol dm^{-3}) in absence of light, which is the most probable one in the oral cavity.

Part II will present data of conductance for the identical systems (i.e., absence and presence of light) and we will discuss the effect of light on both transport properties (diffusion and conductance), and the main conclusions are presented.

12.2 MATERIAL AND METHODS

12.2.1 REAGENTS

KSCN (Merck > 0.99) was used without further purification. The solutions for the diffusion measurements were prepared using Millipore-Q water (specific resistance = 1.82×10^5 Ω m, at 25 °C). All solutions were freshly prepared at 25 °C before each experiment.

12.2.2 CONDUCTOMETRIC TECHNIQUE

The open-ended capillary cell employed, which has previously been used to obtain mutual diffusion coefficients for a wide variety of electrolytes,[8,20] has been described in great detail in previous papers.[20] Basically, it consists of two vertical capillaries, each closed at one end by a platinum electrode, and positioned one above the other with the open ends separated by a distance of about 14 mm. The upper and lower tubes, initially filled with solutions of concentrations 0.75 *c* and 1.25 *c*, respectively, are surrounded with a solution of concentration *c*. This ambient solution is contained in a glass tank (200 × 140 × 60) mm immersed in a thermostat bath at 25 °C. Perspex sheets divide the tank internally and a glass stirrer creates a slow lateral flow of ambient solution across the open ends of the capillaries. Experimental conditions are such that the concentration at each of the open ends is equal to the ambient solution value *c*, that is, the physical length of the capillary tube coincides with the diffusion path. This means that the required boundary conditions described in the literature[20] to solve Fick´s second law of diffusion are applicable. Therefore, the so-called Δl effect[20] is reduced to negligible proportions. In our manually operated apparatus, diffusion is followed by measuring the ratio $w = R_t/R_b$ of resistances R_t and R_b of the upper and lower tubes by an alternating current transformer bridge. In our automatic apparatus, w is measured by a Solartron digital voltmeter (DVM) 7061 with 6 1/2 digits. A power source (Bradley Electronic Model 232) supplies a 30 V sinusoidal signal at 4 kHz (stable to within 0.1 mV) to a potential divider that applies a 250 mV signal to the platinum electrodes in the top and bottom capillaries. By measuring the voltages V' and V'' from top and bottom electrodes to a central electrode at ground potential in a fraction of a second, the DVM calculates w.

In order to measure the differential diffusion coefficient D at a given concentration c, the bulk solution of concentration c is prepared by mixing 1 L of "top" solution with 1 L of "bottom" solution, accurately measured. The glass tank and the two capillaries are filled with c solution, immersed in the thermostat, and allowed to come to thermal equilibrium. The resistance ratio $w = w_\infty$ measured under these conditions (with solutions in both capillaries at concentration c accurately gives the quantity $\tau_\infty = 10^4/(1 + w_\infty)$.

The capillaries are filled with the "top" and "bottom" solutions, which are then allowed to diffuse into the "bulk" solution. Resistance ratio readings are taken at various recorded times, beginning 1000 min after the start of the experiment, to determine the quantity $\tau = 10^4/(1 + w)$ as τ approaches τ_∞. The diffusion coefficient is evaluated using a linear least-squares procedure to fit the data and finally, an iterative process is applied using 20 terms of the expansion series of Fick's second law for the present boundary conditions. The theory developed for the cell has been described previously.[20]

12.3 RESULTS

Mutual diffusion coefficients, D of KSCN in aqueous solutions, from 0.010 to 0.20 mol dm^{-3} were measured in absence of light by using a dark chamber and at 25 °C, and the respective standard deviations[31] of the means are shown in Table 12.1. *D* is the mean value of at least three independent measurements and our uncertainty is not much greater than 1–2%.

TABLE 12.1 Diffusion Coefficients, *D*, of KSCN in Aqueous Solutions at Various Concentrations, c, Measured in the Absence of Light (in Dark Chamber) and at 25 °C

c (mol dm^{-3})	D (10^{-9} m^2 s^{-1})	E_a (10^{-9} m^2 s^{-1})[a]	C_v(%[a])
1×10^{-2}	1.737	0.013	1.5
2×10^{-2}	1.868	0.011	1.2
5×10^{-2}	1.971	0.007	1.2
8×10^{-2}	1.936	0.015	1.5
1×10^{-1}	1.943	0.007	0.7
2×10^{-1}	1.961	0.004	0.4

[a]E_a and C_v represent the error calculated from the standard deviation, considering a confidence limit of 95%, and the coefficient of variation, respectively.

Given the small relative deviations found between the diffusion coefficients obtained by the present method and those obtained by Mitchell et al.,[47],that is, equal to or less than ±2%, it can be considered that such results are in agreement with each other. Contrary to this situation, there are large deviations between our data obtained in presence and absence of light, and obtained by Mitchell et al.,[47] that is, 6–14% approximately. A possible explanation for these discrepancies will be given in the next section.

12.4 INTERPRETATION OF DIFFUSION BEHAVIOR OF KSCN OBTAINED IN THE ABSENCE OF LIGHT

Experimental diffusion coefficients for KSCN solutions when obtained under different conditions, that is, in the presence and absence of light revealed significant deviations among them (approximately 8–12%). A possible interpretation of this phenomenon is that, under those conditions, we may be in the presence of two systems with different structures reflecting thus, different behavior on their diffusion.

In fact, photochemical studies performed with KSCN solutions,[48,49] being subjected to radiation of wavelengths between 180 and 300 nm, reveal the decomposition of the respective anion (eq 12.1), a postulated reaction to be reversed when such solutions are handled in a darkroom. On the other hand, in addition to sulfur formation[50] and cyanide ions, another process occurs simultaneously (eq 12.2), resulting in radical formation $(CNS)_2^-$.[50]

$$CNS^-.H_2O \leftrightharpoons CNS^{-*} \leftrightharpoons CN^- + S \quad (12.1)$$

$$CNS^-.H_2O \leftrightharpoons CNS^-.H_2O^* \leftrightharpoons CNS + e^- \quad (12.2)$$

$$CNS + CNS^- \leftrightharpoons (CNS)_2^-$$

Thus, based on this information and assuming that such processes only occur in solutions prepared in the presence of sunlight, a reasonable explanation is obtained about the different diffusion behavior of this system under the mentioned conditions (in the presence or absence of light). That is, if on one hand there is sulfur formation, eventually such species become responsible for the partial obstruction of ionic movement, on the other hand there are new species $(CNS)2^-$ which, being considerably larger than that of SCN ion, present lower mobility, leading to low diffusion coefficients.

12.5 CONCLUSIONS

Diffusion coefficients were measured for aqueous solutions of KSCN in absence of light and comparing with those previously obtained in the presence of light, providing transport data necessary to model the diffusion for various chemical and pharmaceutical applications.

Our data show that there are significant deviations among the diffusion coefficients measured in those conditions, and in the studied concentration range ($\Delta D \leq 10\%$). Thus, the significant differences lead us to conclude that light plays an important role in the diffusion of SCN species in aqueous solutions. A possible explanation for the diffusion behavior differences for that system in the presence and absence of light is mainly due to the presence of different structures, revealing our method to be sensitive to this fact.

ACKNOWLEDGMENT

The authors are grateful for funding from "The Coimbra Chemistry Centre" which is supported by the Fundação para a Ciência e a Tecnologia (FCT), Portuguese Agency for Scientific Research, through the programmes UID/QUI/UI0313/2019 and COMPETE, and CIROS, Centro de Investigação e Inovação em Ciências Dentárias da FMUC.

KEYWORDS

- **diffusion coefficient**
- **potassium thiocyanate**
- **electrolytes**
- **solutions**
- **transport properties**

REFERENCES

1. Lobo, V. M. M. The Definition of Electrolyte: A Comment from a Reader and the Author's Reply, Port. *Electrochim. Acta* **1997,** *15*, 215.
2. Lobo, V. M. M. The Definition of Electrolyte, Port. *Electrochim. Acta* **1996,** *14*, 27.
3. Robinson, R. A.; Stokes, R.H. *Electrolyte Solutions*, 2nd ed.; Butterworths: London, 1959.
4. Tyrrell, H. J. V.; Harris, K. R. *Diffusion in Liquids*, 2nd ed.; Butterworths: London, 1984.
5. Harned, H. S.; Owen, B. B. *The Physical Chemistry of Electrolytic Solutions*, 3rd ed.; Reinhold Pub. Corp.: New York, 1964.
6. Erdey-Grúz, T. *Transport Phenomena in Aqueous Solutions*; Adam Hilger: London, 1974.
7. Cussler, E. L. *Diffusion: Mass Transfer in Fluid Systems*; Cambridge University Press: Cambridge, 1984.
8. Ribeiro, A. C. F.; Lobo, V. M. M.; Valente, A. J. M.; Cabral, A. M. T. D. P. V.; Veiga, F. J. B.; Fangaia, S. I. G.; Nicolau, P. M. G.; Guerra, F. A. D. R. A.; Esteso, M. A. Transport Properties and Their Impact on Biological Systems. In *Advances in Chemistry Research*; Taylor, J. C., Eds. Nova Science Publishers: New York, Vol. 10, 2011; pp. 379-391.
9. Lobo, V. M. M. *Handbook of Electrolyte Solutions*; Elsevier Sci. Publ.: Amsterdam, 1990.

10. Horvath, A.L. *Handbook of Aqueous Electrolyte Solutions. Physical Properties. Estimation and Correlation Methods*; John Wiley and Sons: New York, 1985.
11. Koike, M.; Fujii, H. The Corrosion Resistance of Pure Titanium in Organic Acids. Biomaterials **2001,** *22*, 2931.
12. Schiff, N.; Grosgogeat, B.; Lissac, M.; Dalard, F. Influence of Fluoride Content and pH on the Corrosion Resitance of Titanium and its Alloys. *Biomaterials* **2002,** *23*, 1995.
13. Alías, J. F. L.; Gomis, J. M.; Anglada, J. M.; Peraire, M. Ion Release from Dental Casting Alloys as Assessed by a Continous Flow System: Nutritional and Toxicological Implications. *Dent. Mater.* **2006,** *22*, 832.
14. Upadhyay, D.; Panchal, M.A.; Dubey, R.S.; Srivastava, V.K. Corrosion of Alloys Used in Dentistry. *Mater. Sci. Eng. A* **2006,** *432*, 1.
15. Rezende, M. C. R. A.; Alves, A. P. R.; Codaro, E. N.; Dutra, C. A. M. Effect of Commercial Mouthwashes on the Corrosion Resitance of Ti-10Mo Experiment Alloy. *J. Mater. Sci: Mater Med.* **2007,** *18*, 149.
16. Rahm, E.; Kunzmann, D.; Döring, H.; Holze, R. Corrosion Stable Nickel and Cobalt-Based Alloys for Dental Applications. *Microchim. Acta* **2007,** *156*, 141.
17. Robin, A.; Meirelis, J. P. Influence of Fluoride Concentration and pH on Corrosion Behavior of Titanium in Artificial Saliva. *J. Appl. Electrochem*. **2007,** 37, 511.
18. Nascimento,M. L.; Mueller, W. D.; Carvalho, A. C.; Tomás, H. Electrochemical Characterization of Cobalt Based Alloys Using the Mini Cell System. *Dent. Mater.* **2007,** *23*, 369.
19. Schultz, C. P.; Ahmed, M. K.; Dawes, C.; Mantsch, H. H. Thiocyanate Levels in Human Saliva: Quantitation by Fourier Transform Infrared Spectroscopy. *Anal. Biochem.* **1996,** 15, 240.
20. Agar, J. N.; Lobo, V. M. M. Measurement of Diffusion Coefficients of Electrolytes by a Modified Open-ended Capillary Method. *J. Chem. Soc., Faraday Trans. I* **1975,** *71*, 1659.
21. Lobo, V. M. M. Mutual Diffusion Coefficients in Aqueous Electrolyte Solutions. *Pure Appl. Chem.* **1993,** 65, 2613.
22. Lobo, V. M. M.; Teixeira, M. H. S. F. Diffusion Coefficients in Aqueous Solutions of Hydrochloric Acid at 25 °C. *Electrochim. Acta* **1979,** *24*, 565.
23. Lobo, V. M. M.; Teixeira, M. H. S. F.; Quaresma, J. L. Diffusion Coefficients in Aqueous Solutions of Potassium Perchlorate at 25 °C. *Electrochim. Acta* **1982,** *27*, 1509.
24. Lobo, V. M. M. Mutual Diffusion Coefficients in Aqueous Electrolyte Solutions. *Pure Appl. Chem.* **1993,** *65*, 2613.
25. Lobo, V. M. M.; Ribeiro, A. C. F.; Veríssimo, L. M. P. Diffusion Coefficients in Aqueous Solutions of Magnesium Nitrate at 298 K. *Ber. Bunsenges Phys. Chem.* **1994,** *98*, 205.
26. Lobo, V. M. M.; Ribeiro, A. C. F.; Veríssimo, L. M. P. Diffusion Coefficients in Aqueous Solutions of Beryllium Sulphate at 298 K. *J. Chem. Eng. Data* **1994,** *39*, 726.
27. Lobo, V. M. M. Diffusion and Thermal Diffusion in Electrolyte Solutions, Port. *Electrochim. Acta* **1995,** *13*, 227.

28. Lobo, V. M. M.; Ribeiro, A. C. F.; Andrade, S. G. C. S. Diffusion Coefficients in Aqueous Solutions of Divalent Electrolytes. *Ber. Buns. Phys. Chem.* **1995,** *99*, 713.
29. Lobo, V. M. M.; Ribeiro, A. C. F.; Valente, A. J. M. Célula de Difusão Condutimétrica de Capilares Abertos – Uma Análise do Método. *Corros. Prot. Mat.* **1995,** *14*, 14.
30. Lobo, V. M. M.; Ribeiro, A. C. F.; Veríssimo, L. M. P. Diffusion Coefficients in Aqueous Solutions of Potassium Chloride at High and Low Concentrations. *J. Mol. Liq.* **1998,** *78*, 139.
31. Ribeiro, A. C. F.; Lobo, V. M. M.; Natividade, J. J. S. Diffusion Coefficients in Aqueous Solutions of Potassium Thiocyanate at 25 °C. *J. Mol. Liq.* **2001,** *94*, 61.
32. Ribeiro, A. C. F.; Lobo, V. M. M.; Azevedo, E. F. G.; Miguel, M. d. G.; Burrows, H. D. Diffusion Coefficients of Sodium Dodecylsulfate in Aqueous Solutions and in Aqueous Solutions of Sucrose. *J. Mol. Liq.* **2001,** *94*, 193.
33. Ribeiro, A. C. F.; Lobo, V. M. M.; Azevedo, E. F. G. Diffusion Coefficients of Ammonium Monovanadate in Aqueous Solutions at 25 °C. *J. Sol. Chem.* **2001,** *30*, 1111.
34. Ribeiro, A. C. F.; Lobo, V. M. M.; Natividade, J. J. S. Diffusion Coefficients in Aqueous Solutions of Cobalt Chloride at 25 °C. *J. Chem. Eng. Data* **2002,** *47*, 539.
35. Lobo, V. M. M.; Valente, A. J. M.; Ribeiro, A. C. F. *Differential Mutual Diffusion Coefficients of Electrolytes Measured by the Open-Ended Conductimetric Capillary Cell: a Review*, *Focus on Chemistry and Biochemistry*; Nova Science Publishers: New York, 2003; p. 15–38.
36. Ribeiro, A. C. F.; Lobo, V. M. M.; Azevedo, E. F. G.; Miguel, M. d. G.; Burrows, H. D. Diffusion Coefficients of Sodium Dodecylsulfate in Aqueous Solutions and in Aqueous Solutions of β-cyclodextrin. *J. Mol. Liq.* **2003,** *102*, 285.
37. Valente, A. J. M.; Ribeiro, A. C. F.; Lobo, V. M. M.; Jiménez, A. Diffusion Coefficients of Lead(II) Nitrate in Nitric Acid Aqueous Solutions at 25 °C. *J. Mol. Liq.* **2004,** *111*, 33.
38. Ribeiro, A. C. F.; Lobo, V. M. M.; Valente, A. J. M.; Azevedo, E. F. G.; Miguel, M.d.G.; Burrows, H. D. Transport Properties of Alkyltrimethylammonium Bromide Surfactants in Aqueous Solutions. *Coll. Polym. Sci.* **2004,** *283*, 277.
39. Ribeiro, A. C. F.; Lobo, V. M. M.; Oliveira, L. R. C.; Burrows, H. D.; Azevedo, E. F. G.; Fangaia, S. I. G.; Nicolau, P. M. G.; Guerra, F. A. D. R. A. Diffusion Coefficients of Chromium Chloride in Aqueous Solutions at 298.15 and 303.15 K. *J. Chem. Eng. Data* **2005,** *50*, 1014.
40. Ribeiro, A. C. F.; Lobo, V. M. M.; Leaist, D. G.; Natividade, J. J. S.; Veríssimo, L. P.; Barros, M. C. F.; Cabral, A. M. T. D. P. V. Binary Diffusion Coefficients for Aqueous Solutions of Lactic Acid. *J. Sol. Chem.* **2005,** *34*, 1009.
41. Ribeiro, A. C. F.; Esteso, M. A.; Lobo, V. M. M.; Valente, A. J. M.; Simões, S. M. N.; Sobral, A. J. F. N.; Burrows, H. D. Diffusion Coefficients of Copper Chloride in Aqueous Solutions at 25 °C and 310.15 K. *J. Chem. Eng. Data* **2005,** *50*, 1986.
42. Ribeiro, A. C. F.; Esteso, M. A.; Lobo, V. M. M.; Valente, A. J. M.; Simões, S. M. N.; Sobral, A. J. F. N.; Ramos, L.; Burrows, H. D.; Amado, A. M.; Amorim da Costa, A. M. Interactions of Copper (II) Chloride with B-Cyclodextrinin Aqueous Solutions at 25 °C. *J. Carbohydr. Chem.* **2006,** *25*, 173.
43. Ribeiro, A. C. F.; Lobo, V. M. M.; Valente, A. J. M.; Simões, S. M. N.; Sobral, A. J. F. N.; Ramos, M. L.; Burrows, H. D. Association between Ammonium Monovanadate

and β-Cyclodextrin as Seen by NMR and Transport Techniques. *Polyhedron* **2006,** *25*, 3581.

44. Ribeiro, A. C. F.; Ortona, O.; Simões, S. M. N.; Santos, C. I. A. V.; Prazeres, P. M. R. A.; Valente, A. J. M.; Loboand, V. M. M.; Burrows, H. D. Binary Mutual Diffusion Coefficients of Aqueous Solutions of Sucrose, Lactose, Glucose and Fructose in the Temperature Range 25 °C to 328.15 K. *J. Chem. Eng. Data* **2006,** 51, 1836–1840.
45. Ribeiro, A. C. F.; Esteso, M. A.; Lobo, V. M. M.; Valente, A. J. M.; Simões, S. M. N.; Sobral, A. J. F. N.; Burrows, H. D. Interactions of Copper (II) Chloride with Sucrose, Glucose and Fructose in Aqueous Solutions. *J. Mol. Struct.* **2007,** 826, 113.
46. Ribeiro, A. C. F.; Natividade, J. J. S.; Esteso, M. A. Differential Mutual Diffusion Coefficients of Binary and Ternary Systems Measured by the Open-Ended Conductimetric Capillary Cell and by the Taylor Technique. *J. Mol. Liquids* **2010,** 156, 58–64.
47. Mitchell, J. P.; Butler, J. B.; Albright, J. G. Measurement of Mutual Diffusion-Coefficients, Densities, Viscosities, and Osmotic Coefficients for the System KSCN-H_2O at 25-Degrees-c. *J. Sol. Chem.* **1992,** *21*, 1115.
48. Dogliotti, L.; Hayon, E. Flash Photolysis Study of Sulphite, Thiocyanate, and Thiosulfate Ions in Solution. *J. Phys. Chem.* **1968,** *72*, 1800.
49. Wells, C. H. J. *Introduction to Molecular Photochemistry*; Chapman and Hall Ltd: London, 1972.
50. Gray, H.B. *Electrons and Chemical Binding*; W.A. Benjamin, Inc.: New York, 1964.

CHAPTER 13

Drug Discovery, Drug-Likeness Screening, and Bioavailability: Development of Drug-Likeness Rule for Natural Products

E. D AHIRE, V. N. SONAWANE, K. R. SURANA, and G. S. TALELE*

Department of Pharmaceutics, Sandip Institute of Pharmaceutical Sciences, Nashik, Maharashtra, India

**Corresponding author. E-mail: swatitalele77@gmail.com*

ABSTRACT

Drug likeness is a qualitative conception applied in drug design for how "drug-like" an element is related to factors such as bioavailability and extensively incorporated into the initial phases of lead and drug discovery. It is projected from the molecular structure earlier that the substance is at least synthesized and tested. A traditional technique to estimate drug likeness is to verify compliance of Lipinski's Rule of Five, which contains the amounts of hydrophilic groups, molecular weight, and hydrophobicity. Methods to recognize drug-like molecules are grounded on their capability to discriminate known drugs from nondrugs in the groups of compounds by associating with one or more of the succeeding extensively available drug databases. There are different databases to categorize drug-like molecules which are based on their capability to discriminate known drugs from nondrugs in the set of compounds and have different methods to assess the drug likeness. The concept of drug likeness has numerous applications in drug discovery.

13.1 INTRODUCTION TO DRUG LIKENESS SCREENING

Drug likeness, a qualitative stuff of compounds consigned by specialists' committee vote, is extensively incorporated into the initial phases of lead and the drug discovery. Numerous medicine contestants are unsuccessful in the clinical trials because of dissimilarity in the effectiveness in contradiction of the proposed drug target.[1] Pharmacokinetic and toxicity problems are responsible for other than half of entire letdown in the clinical trials. Consequently initial part of the virtual screening estimates drug likeness of minor molecules, medicament like molecules reveal favorable absorption, distribution, metabolism, excretion, and toxicological (ADMET) properties.[2] The expression "drug-like" is charming further widespread. According to Walters and Murcko, drug-like molecules are compounds which comprise functional groups and/or have physical parameters consist with the middle-of-the-road of known drugs, and therefore can be concluded as compounds which might be active biologically or might display therapeutic potential.[3,4]

Lipinski describes those molecules as "drug-like," which have appropriately acceptable ADMET properties to survive over the Phase I clinical trials.[5] However, drugs and drug-like molecules are dispersed tremendously meagerly over chemical space, which is projected to contain 10^{40}–10^{100} compounds. For drug properties such as synthetic easily, stability, oral availability, good pharmacokinetic parameters, absence of toxicity, and least addictive potential are of greatest significance. Many of these properties rest on the traditional biological and physicochemical parameters of the compounds; nonetheless, the complex structure of the entire drug compounds makes correlating attempts problematic. One interesting methodology is to study the parameters of the fragments of the entire drug compound.[6] The existing review discovers what makes a compound drug like, the approaches for forecast of drug likeness, besides with notes on currently existing drug-like and nondrug-like databases. To recognize the conception of "drug likeness," it is essential to recognize the common properties present in a drug molecule. Bemis and Murcko have done a broad analysis of the forms of molecules with the assistance of an unassuming graph method that reflects only 2D structures.[7] Agreeing to this methodology any molecule can be dismembered into four parts: ring, linker, side chain, and finally framework as shown in Fig. 13.1.

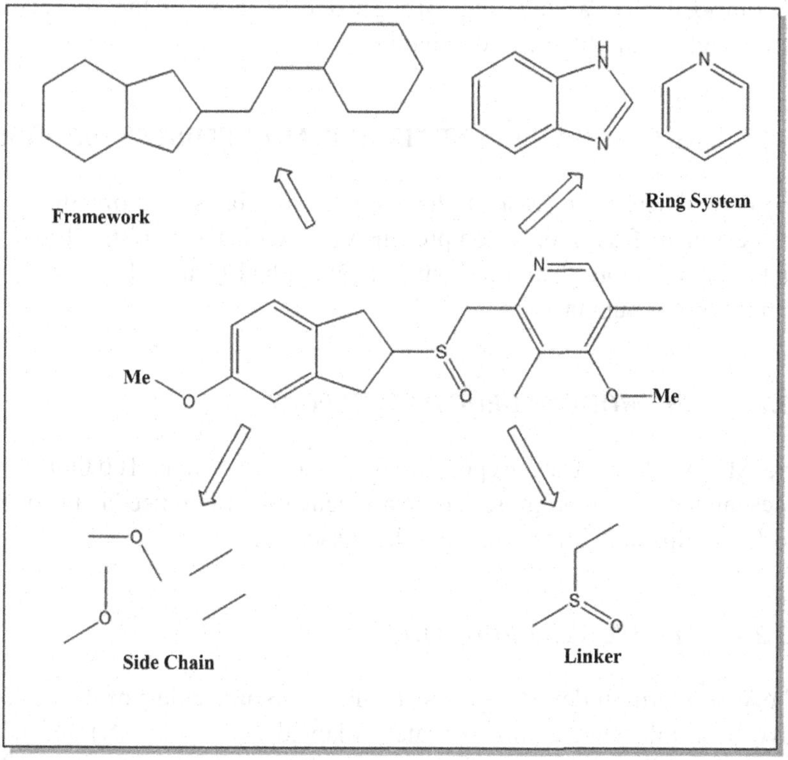

FIGURE 13.1 Hierarchical description of molecules.

Ring scheme is the cyclic portion surrounded by the graph illustration of a molecule and partaking an edge (a linking amid of 2 mol or a bond). For example, omeprazole has benzimidazoline and pyridine ring schemes. Linker molecules form the straight path joining the two rings. On the side chain, atoms are somewhat nonlinker atoms, nonring such as the four side chains in omeprazole, two single-atom side chains, and two-atom side chains.[8] As a final point outline is defined as the amalgamation of ring arrangements and linkers in a wire edge. To regulate what creates a molecule drug-like one necessarily begins by examining the atoms in hand by means of groups of known drugs and with groups of nondrug (or molecules supposed to be nondrug). Approaches to recognize drug-like molecules are grounded on their capability to discriminate recognized drugs from nondrugs in the group of composites by associating through one or more of the succeeding extensively available drug databanks.[9,10]

13.2 THE COMPREHENSIVE MEDICINAL CHEMISTRY DATABASE

The comprehensive medicinal chemistry is plagiaristic from the drug collection in Pergamon's Comprehensive Medicinal Chemistry. The databank contains more than 7000 amalgams, applied or tested as medicinal compounds in human.[11,12]

13.2.1 THE MODERN DRUG DATA REPORT

The Modern Drug Data Report encompasses more than 100,000 medicines launched or in progress. These molecules are referenced in the patent works, symposium arranged, and other bases.[13]

13.2.2 THE WORLD DRUG INDEX

The World Drug Index 1997 consists of 51,596 molecules; of these, 7570 have been allocated a United States Adapted Name (USAN) and 6307 have been allocated as an International Nonproprietary Name (INN); coalescing these gives 8323 exclusive compounds, in which 3515 have initiated in the indication and application arena.[14]

13.2.3 AVAILABLE CHEMICAL DATABASES

The Available Chemical Databases (ACD) is a gathering of more than 300,000 marketed existing compounds. The group of nondrugs is characteristically formed by choosing arbitrary compounds from the existing chemical databanks.[11,15] Using the following kinds of techniques presently measures drug likeness.[16]

13.2.3.1 SIMPLE COUNTING METHOD

Databank assortments of the known drug are characteristically applied to extract information regarding structural characteristics of the active drug molecules. Molecular weight, lipophilicity, and charge are summarized to extract simple counting procedures for applicable explanation of ADMET-associated factors. Simple counting technique includes association of molecular descriptors or characteristics implied to drug likeness. Characteristics such as oral bioavailability or membrane permeability have frequently been interrelated to log *P*, molecular weight (MW) and number of hydrogen bond acceptors, and donors in a compound. The "rule of 5" (RO5) delivers rules for considering if a molecule will be orally bioavailable. The rules were resulting from the investigation of 2245 compounds with a USAN or INN and the entries in the signs and applied arena of the databanks were comprised in the analysis. The hypothesis was that molecules satisfying these criteria had initiate human clinical trials and consequently essential to have influenced many of the necessary features of drugs. The RO5 explains that compounds' presentation less absorption or permeation is more possible to have more than 5 H-bond donors, molecular weight (MWT) over 500, log *P* over 5, and additional 10 H-bond acceptors. Nevertheless, there are abundant examples existing for RO5 defilement amid the available drugs. Mainstream defilements originate from antibiotics, antifungals, vitamins, and cardiac glycoside-like compounds. Still, these class of molecules are orally bioavailable since they hold the groups in which act as substrates for transporters. If a molecule flops the RO5, there is a great possibility that oral activity complications will be encountered, although passing the RO5 is no assurance that a molecule is druglike.[17,18]

13.2.3.2 KNOWLEDGE-BASED METHODS

Knowledge-based approaches are grounded upon the ideas of inherent binding energies and recording of structural fragments. In this technique, mostly functional groups are used to categorize drug and nondrug like compounds grounded on dissimilar scoring functional group fragment. Andrews et al. used a group of 200 drug compounds to originate a group of inherent binding energies for the 10 functional groups as shown in Table 13.1.

TABLE 13.1 Functional Groups Used in the Scoring Scheme Developed by Andrews.

Functional group	Score
Carboxylate	8.2
Phosphate	10
N^+	11.5
N	1.2
OH	2.5
O or S ether	1.1
Halogens	1.3
CO	3.4
C (sp^2)	0.7
C (sp^3)	0.8

The intrinsic binding of small compounds was then projected by summing the intrinsic binding energies and deducting an entropic factor; the technique had been extensively used formerly for the reagent choice rather than drug-likeness extrapolation.[19] The compounds with a score within 2 and 7 were categorized as drugs else they were categorized as nondrugs. Molecules comprising only pharmacophoric group would merely be categorized as drugs if they confined one of the sets of marked by an asterisk in the list of fragments.[12,20]

13.2.3.3 FUNCTIONAL GROUP FILTER

Reactive, toxics, or else inappropriate molecules, like natural product derivatives, are detached using specific filters. Typical reactive functional groups comprise, for example, reactive alkyl halide peroxide, as well as carbazide. Unsuitable leads may comprise crown ethers, disulfide, and aliphatic methylene chain seven or extralong and inappropriate natural products that might include quinones, polyenes, and cycloheximidine derivative removed by means of filters. Screening out the molecules that comprise definite atom sets are related to toxicity delivers a practical and fast way to decrease large databank. Better explanation of toxicity may deliver structure-based technique to evaluate toxicity of the molecule.[17,18]

A diverse method is to recognize functional groups that incline to be unwanted because of chemical reactivity and metabolic ability. Walter et al.

momentarily defined a methodology Rapid Elimination of Swill (REOS) to eradicate objectionable reagent in combinatorial collections. REOS is a hybrid technique that associates some simple counting systems like those in the RO5 with a group of functional groups filter to eliminate the reactive and otherwise underneath sizable moieties.[21] The functional group filters employed in REOS recognize reactive, toxic, and else disagreeable moieties. Preliminary filtering is based on a set of seven property filters. The hydrogen bond donors, acceptors, and electrically charged groups are unwavering using a group of rules alike to those used in the Programmable Atom Typer and Language for Automatic Classification of Atoms in Molecular (PATTY) database established at Merck. Log *P* can be calculated grounded on a diversity of outlines. A web-based boundary marks it trivial to adapt parameters to outfit the requirements of a specific drug-discovery scheme. Examples of the functional group filters engaged by REOS are listed in Table 13.2. In REOS, the functional group filters are identified consuming the SMARTS pattern identical language established at Daylight Chemical Information Systems. SMARTS is an extended version of the Simplified Molecular Input Line Entry System notation established unambiguously for substructure searching. Steps in REOS investigation are as follows: In the first stage, reagents are filtered; reactive and toxic reagents are eradicated while adding to the reagents that evidently will generate a product that interrupts the molecular weight limits. In the succeeding step, reagents tested for compatibility with chemistry, for example, when producing amide, one can make simpler the chemistry by eradicating acids comprising basic amines and amines encompassing acidic functionality.[22] Lastly, the product is filtered keeping in mind the characteristics such as log *P*. This step similarly includes a maximum count cut-off for the functional groups. The most important benefit of SMARTS arrangements is that they are simple ASCII text, which can be simply adapted and used by a diversity of applications. Nevertheless, writing such patterns receipts a bit of rehearsal and representation may not be instantaneously reachable to medicinal chemists.[12,23]

TABLE 13.2 Functional Group Filter Employed by REOS Program.

Functional group	Smart notation
Sulfonyl halide	S(=O)(=O) [F, Cl, Br, I]
Acid halide	C(=O) [Cl, Br, I]
Peroxide	OO
Aldehyde	[HC]=O

13.2.3.4 TOPOLOGICAL DRUG CLASSIFICATION

It is usually expected that compound those having the structural resemblance with known drug may display drug-like characteristics themselves, like oral bioavailability, low toxicity, good membrane permeability, and better metabolic stability. The first part is simulated neural networks and decision trees very fast filter devices in virtual screening approaches. Data are also composed to find structural motifs and pharmacophore properties of small compounds that illustrate drugs. The analysis of virtual libraries rendering to the occurrence or absence of drug-like framework, side chain or structural motifs can be applied for virtual screening.[24]

13.2.3.5 PHARMACOPHORE FILTER

An unassuming pharmacophore filter has been announced in recent times. It is grounded on the hypothesis that drug-like compounds should comprise at least two diverse pharmacophore groups; four functional motifs have been recognized that promise hydrogen bonding competence that is important for the definite collaboration of the drug molecules with its biological targets. These motifs can be united to functional groups that are also mentioned here as pharmacophore points; they comprise amine amide, alcohol, ketone, sulfone, sulfonamide, carboxylic acid carbamate, guanidine, amidine, urea, ester, and so on.

13.2.3.6 MULTIPROPERTY OPTIMIZATION

When scheming a combinatorial library, drug-like behavior denotes only as a small number of characteristics, which must be improved; it may also be essential to improve diversity, potency, selectivity, or a number of other characteristics. Simultaneous optimization of numerous characteristics of combinatorial library includes assortment of random subset of reagents, building of a virtual library of molecules from these reagents, calculation of the characteristics of combinatorial products, constructing amendments to the reagent subset and uncomplaining the variations if they increase the like characteristics of the library. This progression is repetitive until a preprogrammed stopping situation has been obtained.[25–27]

13.2.3.7 CHEMISTRY SPACE METHOD

The elementary hypothesis of these approaches is that drugs will have a tendency to own different standards for assured characteristics and as a result will be presented to be separate from nondrugs when examined in multidimensional space. A chemistry space is characteristically defined by calculating an amount of descriptors for molecule and exhausting the descriptor standards as points in a multidimensional space. For example, let us consider that we have calculated MW, log *P*, and no of H-bond donors for a group of molecules. These three descriptor standards can then be applied to describe a point in a three-dimensional space which characterizes respectively molecule. Compounds are then allotted a drug-likeness index amongst 0 and 100% over an assessment of the descriptor vector for an assumed compound with the cluster center.[28]

13.3 APPLICATION OF DRUG LIKENESS

13.3.1 COMPOUNDS AND SCREENING SET SELECTION

One significant application of these methods is in the context of molecules and screening set assortment. For example, methods such as those previously designated can be used to filter a set of molecules from an exterior supplier prior to procurements. The genetic algorithm-based technique applied to the assortment of molecules from the corporate database for creation of a screening group. Profiling by means of the RO5 and PSA measures can also deliver a valued sign of the likely absorption properties of a combinatorial library or screening group.[12]

13.3.2 COMBINATORIAL LIBRARY DESIGN

In addition to basic profiling libraries, this methodology is one-step advanced and has been useful to the enterprise of combinatorial libraries. With one example, a chemist had designated reagents for a combinatorial library (LIB1) in an oral drug-discovery program to augment parameters such as MW and *C*log *P* in an estimated way. A follow-up library (LIB2) was premeditated to improved PSA and RO5 standards much more thoroughly with reagent choice being accomplished by a Monte Carlo search technique. Mutually libraries were consequently tested in a Caco-2 monolayer absorption scheme and both of the considered libraries exposed much-enhanced absorption. These results presented the additional significance of quantities like PSA in compound (library) design in adding to extra-old-style computed descriptors such as *C*log *P* and MW.[29]

13.3.3 VIRTUAL SCREENING OF CHEMICAL DATABASES

Drug likeness is applied as one of the filters in the virtual screening of chemical databases with the persistence to screen in those compounds, which have characteristics similar to that of a known compound. The Rule of Five is applied as a prime filter for selection the chemical databanks, which are adapted in accordance by means of the potent molecule, to accomplished well-organized probing technique to produce only those compounds having drug-like characteristics.[12,30]

13.4 CONCEPT OF PHARMACOPHORE MAPPING

The concept of pharmacophore mapping is first familiarized in 1990 by "Paul Herilich." A pharmacophore is an abstract description of compound properties, which is required for molecular acknowledgment of a ligand by a biological macromolecule. It is the progression of deriving a 3D pharmacophore. A pharmacophore is a group of properties composed with their comparative spatial location that is assumed to be proficient of collaboration with a particular biological target such as hydrogen bond donors and acceptors, positively and negatively charged groups, hydrophobic areas, and aromatic rings.[31]

A pharmacophore is a demonstration of widespread molecular properties such as

- 3D (hydrophobic groups, charged/ionizable groups, hydrogen bond donors/acceptors)
- 2D (substructures)
- 1D (physical or biological)
- Characteristics that are measured to be accountable for a preferred biological activity.

Pharmacophore mapping is the definition and assignment of pharmacophoric properties and the alignment techniques applied to overlay 3D.

Two slightly distinct procedures:

1. That substructure of a molecule that is accountable for its pharmacological activity (cf., chromophore)
2. A group of geometrical restraints amongst precise functional groups that permit the compounds to have biological activity

The procedure of originating pharmacophore is known as pharmacophore mapping. Figure 13.2 represents the process of outcome drug by design.

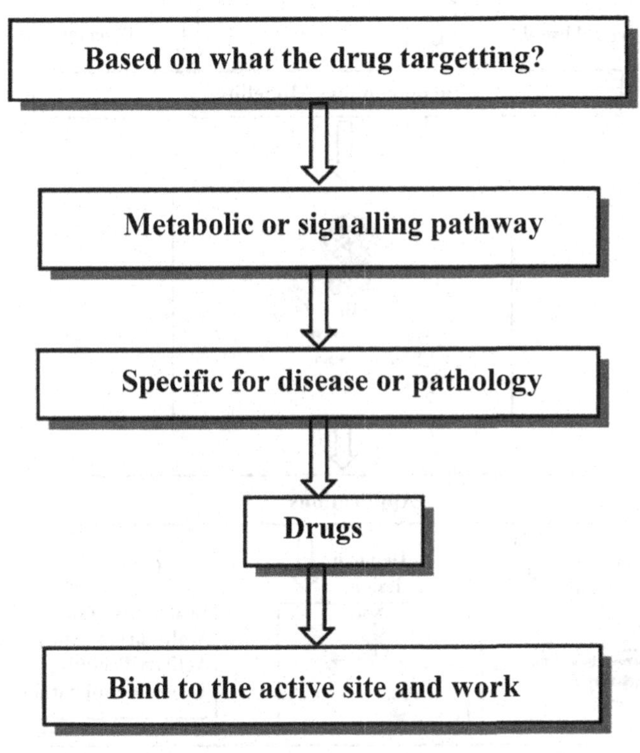

FIGURE 13.2 The process of outcome drug by design.

Pharmacophore mapping consist of three steps:

1. Recognizing common binding element that is accountable for the biological activity
2. Producing probable conformations that active compound may implement
3. Responsible for the 3D association between pharmacophore elements in each conformation produced

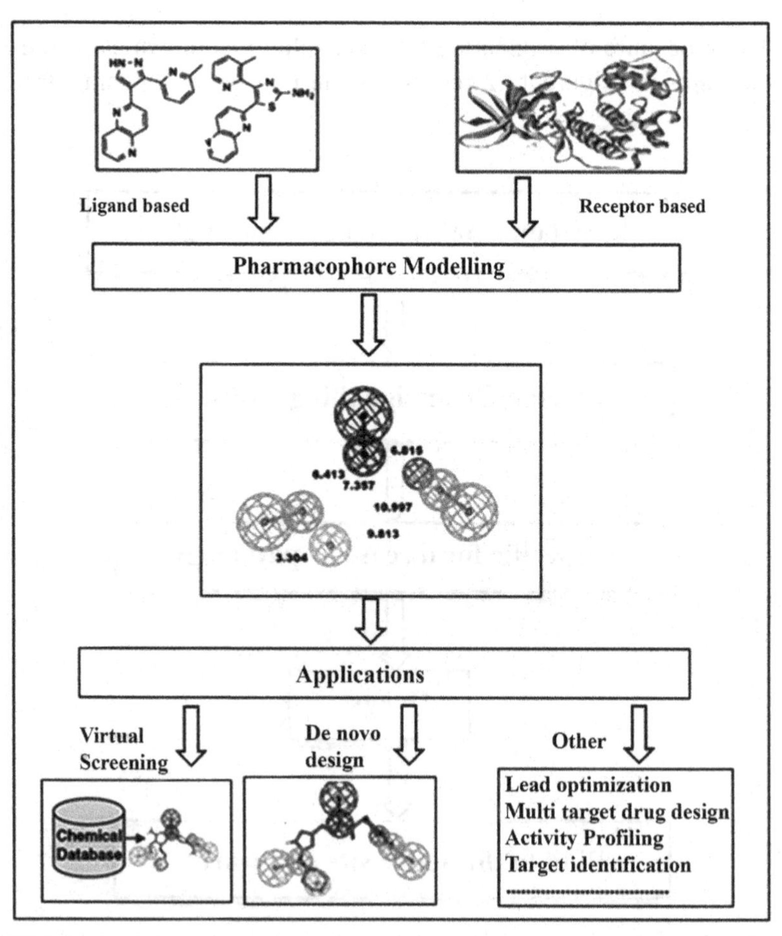

FIGURE 13.3 Schematic representation of pharmacophore modeling.

Typically pharmacophore grounded search is done in two steps:

1. The software checks whether the compound has the atom types or functional groups essential by the pharmacophore.
2. It checks whether the spatial arrangement of this element matches the query.

Flexible 3D searches recognized a greater number of hits than rigid searches do, although flexible searches are more often overwhelming than rigid ones.

There are two key methodologies for comprising conformational flexibility into the search:

1. One is to produce a user-defined number of demonstrative conformation for each compound when the databank is generated.
2. The other is to produce conformation throughout the search.

Pharmacophore model delivers authoritative filter devices for virtual screening even in case where the protein production is not existing, pharmacophore filter is abundant quicker than docking methodologies and consequently significantly decrease the number of molecule exposed to the additional luxurious docking presentation. An alternative exciting feature of pharmacophore in virtual screening is 3D pharmacophore multiplicity.[32]

13.4.1 2D PHARMACOPHORE SEARCHING

Searching of 2D database is of abundant significance for quickening the drug discovery. Different approaches are followed to search a 2D databank to recognize the molecules of the attention. Substructure search recognizes bigger compounds that comprise consumer express query, nevertheless, of the surroundings in which the query substructure happens. Biochemical data available from these molecules can be applied for producing structure–activity relationship (SAR), even formerly synthetic plans are prepared for lead optimization. In distinction, superstructure searches are applied to discover smaller compounds that are surrounded in the query. One problem that can arise from substructure search is that the number of the molecule recognized can reach into the thousands. One resolution of this problem is gathering of the molecules grounded on similarity between molecules in the databank and in the query. Elsewhere, structure similarity and activity similarity have also been the subject of numerous studies.

Resemblance search can be united with substructure for preventive the number of compounds designated. Malleable searches are used to recognize the molecules that vary from the query structure in user-specified methods.[33]

13.4.2 3D PHARMACOPHORE SEARCHING

13.4.2.1 LIGAND-BASED PHARMACOPHORE GENERATION

Ligand-established pharmacophores are usually applied when crystallographic; solution structure or molded structure of protein cannot be achieved. When a group of active molecules is known and it is assumed that all the molecules bind in the same way to the protein, then mutual group should work together with the same protein residue. Therefore, a pharmacophore taking this molecule's feature should be capable to categorize from a database novel molecules that bind to the similar site of the protein as the known molecules do.[34,35]

13.4.3 MANUAL PHARMACOPHORE GENERATION

Manual pharmacophore generation is applied when there is a stress-free technique to recognize the common property in a group of active compounds and/or there is experimental indication that similar functional groups should be extant in the ligand for good activity, for example, the development of a pharmacophore model for dopamine transporter inhibitor. Pharmacophores should also have near flexibility built-in, thus explaining the procedure of distance ranges.[36]

13.4.4 AUTOMATIC PHARMACOPHORE GENERATION

Pharmacophore creation via conformational analysis and manual arrangement is a very time-taking task, specifically when the list of the active ligands is large and the elements of the pharmacophore model are not understandable. There are numerous programs such as Hip Hop, Hypogen, Disco, Gaps, flo, APEX, and ROCS that can spontaneously produce potential pharmacophore from a list of known inhibitors. The presentation of these programs in automated pharmacophore creation varies reliantly on the training set. All these programs use algorithms that recognized the common pharmacophore properties in the training-set compounds; they score utility to rank the recognized pharmacophores.[37]

13.4.5 RECEPTOR-BASED PHARMACOPHORE GENERATION

If the 3D structure of a receptor is well known, a pharmacophore model will be grounded on the receptor's active site. Biochemical data are used to detect the key residue that is significant for substrate and/or inhibiting binding. This evidence can be used for necessary pharmacophores directing the area distinct by significant residue or for selecting among pharmacophore produced by automatic program. This can importantly advance the chance of discovery of small compounds that inhibit the protein since the search is concentrated on an area of the binding side that is vital for binding substrate and inhibitors.[38]

13.5 PHARMACOPHORE-MAPPING SOFTWARE

Discovery studio

- Window ® and Linux® based protein modeling software.[39]
- Produced by Accelrys Software Company.
- Easy to use interface.

Examples of the programs that execute pharmacophore-grounded searches are 3D search UNITY, MACCS-3D, and ROCS. ROCS is used as shape-based superposition for identifying the compound that has similar shape.[40,41]

13.6 CONCLUSIONS

Drug likeness is a key concept in drug discovery and one fascinating methodology is to study the properties of the fragments of whole-drug compound. It discovers what makes a compound drug like, the methods for prediction of drug likeness, along with notes on presently available drug-like and nondrug-like databases. To appreciate the concept of "drug-likeness," it is essential to recognize the common properties present in a drug molecule. Databases and methods are available for the assessment of drug-likeness properties.

KEYWORDS

- **drug-like**
- **drug discovery**
- **drug design**

REFERENCES

1. Zuegg, J.; Cooper, M. A. Drug-Likeness and Increased Hydrophobicity of Commercially Available Compound Libraries for Drug Screening. *Curr. Topics Med. Chem.* **2012,** *12* (14), 1500–1513.
2. Proudfoot, J. R. Drugs, Leads, and Drug-Likeness: An Analysis of Some Recently Launched Drugs. *Bioorganic Med. Chem. Lett.* **2002,** *12* (12), 1647–1650.
3. Jhoti, H.; Leach, A. R. *Structure-based Drug Discovery*; Springer: Germany, 2007.
4. Clark, D. E.; Pickett, S. D. Computational Methods for the Prediction of 'Drug-Likeness'. *Drug Discov. Today* **2000,** *5* (2), 49–58.
5. Giménez, B.; et al. Evaluation of Blockbuster Drugs under the Rule-of-Five. *Die Pharm. Int. J. Pharm. Sci.* **2010,** *65* (2), 148–152.
6. Lipinski, C. A. Lead- and Drug-Like Compounds: The Rule-of-Five Revolution. *Drug Discov. Today: Technol.* **2004,** *1* (4), 337–341.
7. Kubinyi, H.; Mannhold, R.; Timmerman, H. *Virtual Screening for Bioactive Molecules*; John Wiley & Sons: Hoboken, NJ, 2008; Vol. 10.
8. Pollastri, M. P. Overview on the Rule of Five. *Curr. Protocols Pharmacol.* **2010,** *49* (1), 9.12.1–9.12.8.
9. Baell, J.; et al. Ask the Experts: Past, Present and Future of the Rule of Five. *Fut. Med. Chem.* **2013,** *5* (7), 745–752.
10. Zhang, M. Q.; Wilkinson, B. Drug Discovery Beyond The 'Rule-of-Five'. *Curr. Opin. Biotechnol.* **2007,** *18*(6), 478–488.
11. Cummins, D. J.; et al. Molecular Diversity in Chemical Databases: Comparison of Medicinal Chemistry Knowledge Bases and Databases of Commercially Available Compounds. *J. Chem. Inform. Comp. Sci.* **1996,** *36* (4), 750–763.
12. Kadam, R.; Roy, N. Recent Trends in Drug-likeness Prediction: A Comprehensive Review of in Silico Methods. *Indian J. Pharm. Sci.* **2007,** *69* (5), 609.
13. Kong, D.-X.; Li, X.-J.; Zhang, H.-Y. Where Is the Hope for Drug Discovery? Let History Tell the Future. *Drug Discovery Today* **2009,** *14* (3–4), 115–119.

14. Walters, W. P.; Murcko, A. A.; Murcko, M. A. Recognizing Molecules with Drug-like Properties. *Curr. Opin. Chem. Biol.* **1999,** *3* (4), 384–387.
15. Kenny, P. W.; Sadowski, J. Structure Modification in Chemical Databases. *Chemoinf. Drug Discov.* **2005,** *23*, 271–285.
16. Bemis, G. W.; Murcko, M. A. The Properties of Known Drugs. 1. Molecular Frameworks. *J. Med. Chem.* **1996,** *39* (15), 2887–2893.
17. Walters, W. P.; Murcko, M. A. *Prediction of 'Drug-likeness'. Adv. Drug Delivery Rev.* **2002,** *54* (3), 255–271.
18. Muegge, I. Selection Criteria For Drug-Like Compounds. *Med. Res. Rev.* **2003,** *23* (3), 302–321.
19. Andrews, P.; Craik, D.; Martin, J. Functional Group Contributions to Drug-Receptor Interactions. *J. Med. Chem.* **1984,** *27* (12), 1648–1657.
20. Ghose, A. K.; et al. Knowledge-based Chemoinformatic Approaches to Drug Discovery. *Drug Discov. Today* **2006,** *11* (23–24), 1107–1114.
21. Walters, W. P.; Stahl, M. T.; Murcko, M. A. Virtual Screening—An Overview. *Drug Discov. Today* **1998,** *3* (4), 160–178.
22. Zhang, S.-N.; Li, X.-Z.; Yang, X.-Y. Drug-Likeness Prediction of Chemical Constituents Isolated from Chinese Materia Medica Ciwujia. *J. Ethnopharmacol.* **2017,** *198*, 131–138.
23. Turabekova, M. A.; et al. Aconitum and Delphinium Alkaloids: "Drug-Likeness" Descriptors Related to Toxic Mode of Action. *Environ. Toxicol. Pharmacol.* **2008,** *25* (3), 310–320.
24. Galvez, J.; et al. Topological Approach to Drug Design. *J. Chem. Inf. Comp. Sci.* **1995,** *35* (2), 272–284.
25. Gillet, V. J.; et al. Selecting Combinatorial Libraries to Optimize Diversity and Physical Properties. *J. Chem. Inf. Comp. Sci.* **1999,** *39* (1), 169–177.
26. Zheng, W.; et al. PICCOLO: A Tool for Combinatorial Library Design via Multicriterion Optimization. *Biocomputing* **1999,** 2000, 588–599.
27. Press, W. H.; et al. *FORTRAN Numerical Recipes*; Cambridge University Press: England, 1996; Vol. 1.
28. Agrafiotis, D. K.; Myslik, J. C.; Salemme, F. R. Advances in Diversity Profiling and Combinatorial Series Design. *Mol. Divers.* **1998,** *4* (1), 1–22.
29. Brown, R. D.; Hassan, M.; Waldman, M. Combinatorial Library Design for Diversity, Cost Efficiency, and Drug-Like Character. *J. Mol. Graph. Model.* **2000,** *18* (4–5), 427–437.
30. Shoichet, B. K. Virtual Screening of Chemical Libraries. *Nature* **2004,** *432* (7019), 862.
31. Debnath, A. K. Pharmacophore Mapping of a Series of 2,4-diamino-5-deazapteridine Inhibitors of *Mycobacterium avium* Complex Dihydrofolate Reductase. *J. Med. Chem.* **2002,** *45* (1), 41–53.
32. Guner, O. History and Evolution of the Pharmacophore Concept in Computer-Aided Drug Design. *Curr. Topics Med. Chem.* **2002,** *2* (12), 1321–1332.
33. Koes, D. R.; Camacho, C. J. Pharmer: Efficient and Exact Pharmacophore Search. *J. Chem. Inf. Model.* **2011,** *51* (6), 1307–1314.
34. Fang, X.; Wang, S. A Web-based 3D-Database Pharmacophore Searching Tool for Drug Discovery. *J. Chem. Inf. Comp. Sci.* **2002,** *42* (2), 192–198.

35. Mason, J.; Good, A.; Martin, E. 3-D Pharmacophores in Drug Discovery. *Curr. Pharm. Des.* **2001,** *7* (7), 567–597.
36. Güner, O. F. Manual Pharmacophore Generation. *Pharmacophore Perception, Development, and Use in Drug Design*; Internat'l University Line, 2000, Vol. 2, 17.
37. Steindl, T.; Langer, T. Influenza Virus Neuraminidase Inhibitors: Generation and Comparison of Structure-based and Common Feature Pharmacophore Hypotheses and Their Application in Virtual Screening. *J. Chem. Inf. Comp. Sci.* **2004,** *44* (5), 1849–1856.
38. Güner, O. F. *Pharmacophore Perception, Development, and Use in Drug Design*; Internat'l University Line, 2000; Vol. 2.
39. Pluhackova, K.; Wassenaar, T. A.; Böckmann, R. A. Molecular Dynamics Simulations of Membrane Proteins. *Membrane Biogenesis*; Springer: Berlin-Heidelberg, 2013; pp 85–101.
40. Liu, X.; et al. PharmMapper Server: A Web Server for Potential Drug Target Identification Using Pharmacophore Mapping Approach. *Nucleic Acids Res.* **2010,** *38* (Suppl_2), W609–W614.
41. Mannhold, R.; Kubinyi, H.; Folkers, G. *Pharmacophores and Pharmacophore Searches*; John Wiley & Sons: Hoboken, NJ, 2006; Vol. 32.

CHAPTER 14

Biomolecular and Molecular Docking: A Modern Tool in Drug Discovery and Virtual Screening of Natural Products

KHEMCHAND R. SURANA[1], EKNATH D. AHIRE[1], VIJAYRAJ N. SONAWANE[1], and SWATI G. TALELE[2*]

[1]*Divine College of Pharmacy, Nampur Road, Satna, India*

[2]*Sandip Institute of Pharmaceutical Sciences, Mahiravani, Nasik, India*

[*]*Corresponding author. E-mail: swatitalele77@gmail.com*

ABSTRACT

Molecular docking is a computational tool of the molecules of complexes molded by different interactions of molecules. The aim of molecular docking is to predict the 3D structure. Molecular docking shows a significant role in the coherent design of drugs. To achieve an optimized conformation for both the protein and ligand, and relative orientation amongst protein and ligand, several types of docking are used often such as rigid docking, flexible docking, and full flexible docking so that the free energy of the overall system is minimized. De-novo drug design is a process in which the 3D structure of receptor is used to design newer molecules. It involves structural determination of the lead target complexes and lead modifications using molecular modeling tools.

14.1 INTRODUCTION TO MOLECULAR DOCKING

The process of two or more interacting molecules forming the structure of complexes using computational modeling is known as molecular docking.

Three-dimensional structure predictions are the goal of molecular docking. In the designing of drugs, docking plays an important role. The intension of molecular docking is to get an effective conformation for both the protein and ligand, and orientation which is relative to protein and ligand so that the free energy of the complete system is minimized. In promoting basic biomolecular events, such as enzyme–substrate, drug–protein, and drug–nucleic acid interaction, molecular recognition plays a very important role (Fig. 14.1).[1,2]

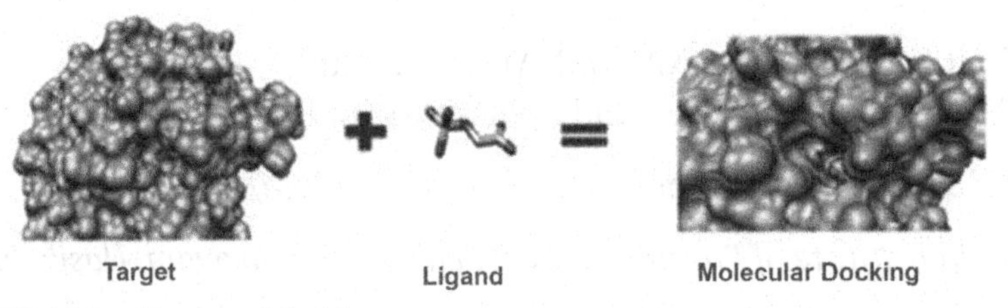

FIGURE 14.1 Process of docking.

Molecular docking is in-silico procedure, which predicts the placement of small molecules or ligands within the active site of their target protein (receptor). It is mainly used to accurate determination of most constructive binding modes and bioaffinities of ligands with their receptor, presently it has been broadly applied to virtual screening for the optimization of the lead compounds. Molecular docking methodology consists of mainly three goals which are linked to each other like[3,4]:

- Binding pose prediction
- Bioaffinity
- Virtual screening

In the molecular docking method, the basic tools are search algorithm and scoring functions for creating and analyzing conformations of the ligand.[5,6]

14.2 BASIC CONCEPT

Docking is the formation of nondent protein–ligand complexes. Suppose the structures of a protein and ligand are given and the prediction of structure of the resulting complex is the task. This is the docking problem. Because the native geometry of the complex can usually be considered to reflect the global minimum of the binding free energy, actually docking is an energy-optimization problem. Therefore, heuristic approximations are necessary to render the problem tractable within a time. The development of docking procedures is therefore concerned with making the correct assumptions and finding satisfactory simplifications that still provide an enough accurate and predictive model for protein–ligand interactions. To fit the protein target, the former procedure designs new ligands while the latter is used to fix on whether present compounds possess a good chemical and steric complementarily to the particular protein (Fig. 14.2).[7]

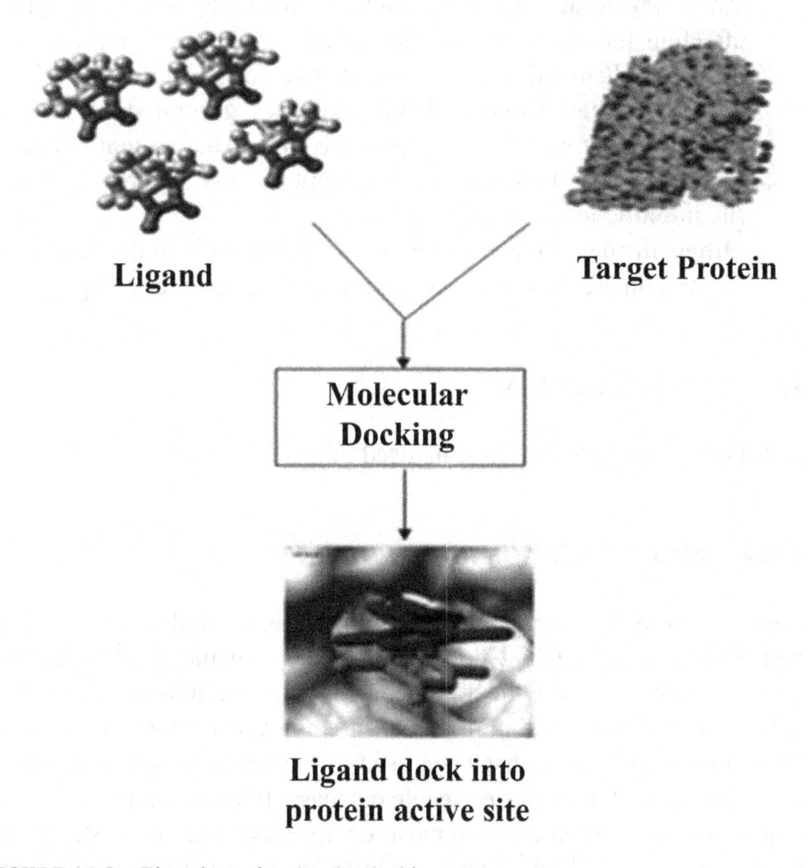

FIGURE 14.2 Flowchart of molecular docking.

14.3 DIFFERENT TYPES OF INTERACTIONS

Interactions among particles can be defined as a consequence of forces linking the molecules contained by the particles. These forces are separated into four categories[8]:

- **Electrostatic forces:** Forces having electrostatic origin caused by the charges residing in the matter. The most common interactions are charge–charge, charge–dipole, and dipole–dipole.
- **Electrodynamics forces:** The usually known is the Van der Waals interactions.
- **Steric forces:** Steric forces are generated when atoms in different molecules come into close contact with each other and initiate affecting the reactivity of each other. The generated forces can influence chemical reactions and the free energy of a system.
- **Solvent-related forces:** These are forces generated because of chemical reactions connecting the solvent and the protein or ligand. Examples are hydrophobic interactions and hydrogen bonds (hydrophilic interactions).
- **Other physical factors:** Conformational changes in the ligand and the protein are frequently necessary for successful docking.

14.4 TYPES OF DOCKING

The following are types of docking used often.

14.4.1 LOCK AND KEY OR RIGID DOCKING

In rigid docking, the internal geometry of the ligand and receptor is kept fixed while docking. Rigid docking, where a suitable position for the ligand in receptor environment is obtained while maintaining its rigidity. Rigid-body docking simulation has been employed for virtual-screening initiatives; this procedure has been used as the fastest way to execute an initial screening of a minute molecule database. It has comparatively high accuracy, when compared in opposition to crystallographic structures. This accuracy is still higher when we introduced an analysis of the finest results using an empirical scoring function for the finest results obtained with rigid-body docking simulations. Generally, flexible docking or scoring functions have been used for applying an extraspecific refinement and lead optimization after initial rigid-body docking method, since these procedures demand for computational power and CPU time.[9–11]

14.4.2 INDUCED-FIT OR FLEXIBLE DOCKING

According to this model, the ligand and side chain of the protein is kept flexible and the energy for dissimilar conformations of the ligand fitting into the protein is determined by the energy for dissimilar conformations. In induced-fit docking, the main chain is used to move to integrate the conformational changes of the protein after ligand binding. Although it is time-consuming and computationally costly, this method can evaluate many different feasible conformations which make it more in-depth and possibly simulate real-life incident and hence trustworthy. Flexible docking somewhere a favored geometry for receptor–ligand interaction is obtained by altering internal torsions of ligand into the active site, whereas receptor remains unchanged. Flexible docking procedures can consider numerous possible conformations of receptor or ligand, as well as at the same time for both the molecules at an elevated computational time cost.[12–14]

14.4.3 FULL-FLEXIBLE DOCKING

However, the ligand is flexed using its torsion angles by the side chain of active site residues (selected active site residues by a user-specified radius around the ligand) are flexed. One key portion of molecular modeling is calculating the energy of conformations and interactions using procedures ranging from quantum mechanics to solely empirical energy functions. Molecular docking energy evaluations are typically carried out with the help of a scoring function. Developing these scoring functions is the most important challenge in drug design. Scoring function is responsible for the accuracy of geometric modeling to obtain right docking. Scoring functions are based usually on force fields that designed to simulate the function of proteins. To incorporate terms such as salvation and entropy, some scoring functions, were used in molecular docking and have been adapted. To specifically select those that show an elevated affinity and to rapidly screen millions of possible compounds that fit a particular receptor is the challenge of the lead-generation phase of the receptor–ligand docking approach. The selected set of ligands can then be screened further by more involved computational procedures, such as free-energy perturbation theory or directly in assays. Many more techniques have been planned that address particular parts of this challenge. Under the theory of both rigid ligands and a rigid protein particular ligand can fit into the receptor pocket.[15,16] This problem allows an approach which is enumerative; because there are just six degrees of freedom that completely clarify the comparative position of the ligand relating to the receptor. Sconuch techniques are considerably fast. Possible geometries can be scored by force field, knowledgeable or empirical-based procedures.[17,18] (MDS allows user to select different intermolecular interactions, namely, steric, electrostatic.)

14.5 MAJOR STEPS IN MECHANICS OF MOLECULAR DOCKING

Molecular docking is the procedure where the intermolecular interaction between two molecules is considered in silico. In this procedure, the protein receptor is a macromolecule. Ligand molecule, which is act as an inhibitor, is the micromolecule. So, the docking procedure involves the following steps[19–21]:

14.5.1 STEP I: PREPARATION OF PROTEIN

Three-dimensional arrangement of the protein should be retrieved from protein data bank (PDB); subsequent to that the retrieved structure should be preprocessed. This should admit taking away of the water molecules from the cavity, filling the absent residues, stabilizing the charges, generating the side chains and so on according to the parameters available.[20]

14.5.2 STEP II: ACTIVE SITE PREDICTION

Subsequent to the preparation of protein, the active site of protein should be predicted. The receptor might have lots of active sites, merely one of the concern should be chosen out. Generally, the heteroatoms and water molecules are removed if present.[20]

14.5.3 *STEP III: LIGAND PREPARATION*

Ligands can be retrieved from a number of databases such PubChem, zinc or can be sketched with Chem sketch tool. While selecting out the ligands, the Lipinski's Rule of Five is to be used. Lipinski rule of D assists in discerning with nondrug-like and drug-like candidates. It promises more chance of success or not a success due to drug similarities for molecules abiding by with two or more than of the complying rules. For choice of a ligand allowing to the Lipinski's rule:

1. Hydrogen bond donors which are less than five
2. Hydrogen bond acceptors which are less than 10
3. Molecular mass less than 500 Da
4. High lipophilicity (expressed as log *P* not over 5)
5. Molar refractivity to be between 40 and 130

14.5.4 *STEP IV: DOCKING*

Ligand is docked adjacent to the protein and the relations are analyzed. The scoring function gives score on the basis of finest docked ligand complex is picked out.[20]

14.6 HOMOLOGY MODELING

In the lack of experimental structures, computational procedures are used to guess the 3D structure of target proteins. Comparative modeling is used to guess target structure based on a template with a similar order, leveraging that protein structure is improved conserved than sequence, that is, proteins with same sequences have same structures. The template and target proteins share the same evolutionary source is a particular type of comparative modeling, which is called as homology modeling (Fig. 14.3).[22] Comparative modeling involves the following steps:

1. Identification of related proteins to provide as template structures.
2. Sequence arrangement of the target and template proteins.
3. Copying coordinates for confidently associated regions.

4. Constructing missing atom coordinates of targeted structure.
5. Model evaluation and refinement.

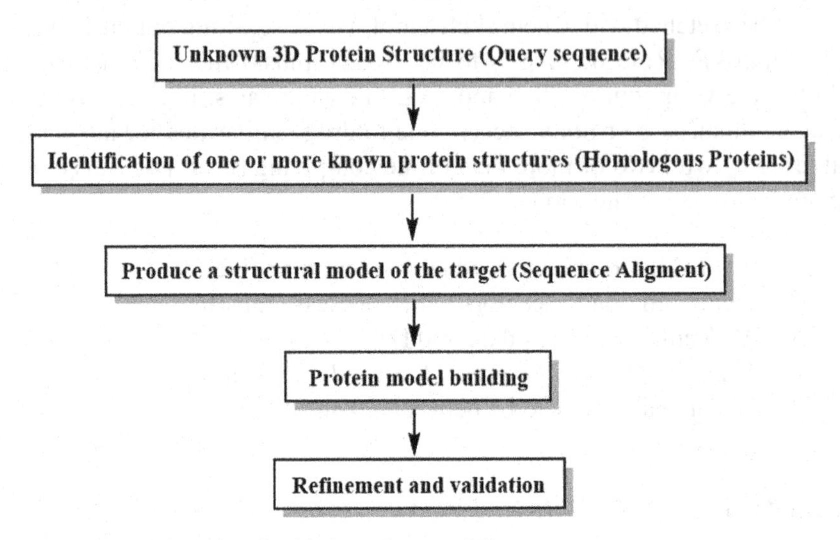

FIGURE 14.3 Steps involved in homology modeling.

Many computer programs and web servers be present that automate the homology modeling process, for example, PSIPRED and MODELLER. Most important goal of structural biology includes formation of protein–ligand complexes, in which the protein molecules act energetically in course of binding. Thus, perceptive of protein–ligand relations will be very significant for structure-based drug design. Absence of understanding of 3D structures has hindered efforts to understand the binding specificities of ligands with protein.[23] Homology modeling is quickly suitable the procedure of choice for obtaining 3D coordinates of proteins with growth in modeling software and increasing in the number of known protein structures. Homology modeling is a symbol of the resemblance of environmental residues at topologically subsequent positions in the reference-based proteins. In the lack of experimental data, model building on the basis of a well-known 3D structure of a homologous protein is the only reliable procedure to obtain the structural information. The awareness of the 3D structures of proteins provides very useful insights into the molecular basis of their functions (Fig. 14.4).[23,24]

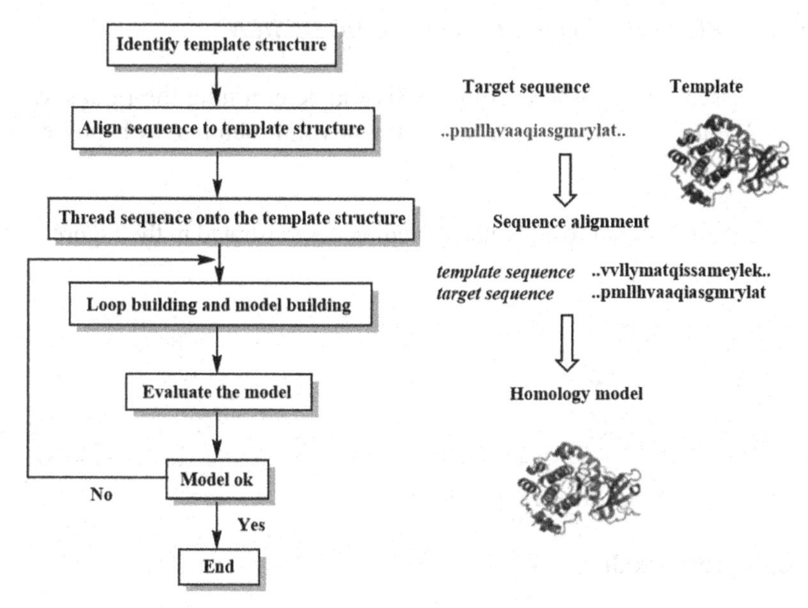

FIGURE 14.4 Steps involved on homology model building process.

14.7 DE-NOVO DRUG DESIGN

De novo means start afresh, from the scratch from the beginning. It is a procedure in which the 3D structure of receptor is utilized to design new molecules. Lead targeted complexes and lead modifications using molecular modeling tools are involved in structural determination. Information present about target receptor but no existing leads that can interact.[25] Bioactive compounds are designed by incremental creation of a ligand model within a model of the enzyme or receptor active site. X-ray or NMR data are used to identify the structure. De-novo design is the use of docking programs design of new lead structures that in shape of a particular target site. Flexible molecules are better than rigid molecules.[26,27] It is pointless that molecular designing is difficult or not possible to synthesize. Likewise, it is pointless designing molecules that require to adopt an unstable conformation to bind. Consideration of the energy losses involving in water desolvation is to be taken into account. There may be suitable differences in structure linking receptors and enzymes from unlike species. This is important if the structure of the binding site used for de-novo design is based on a protein that is not human origin.[28,29]

14.7.1 *PRINCIPLES OF DE-NOVO DRUG DESIGN*

- Assembling possible compounds and determining their quality.
- Identifying the sample space for novel structures with properties like drug.[30]

The protein structure and related binding demonstrated in the Figure 14.5.

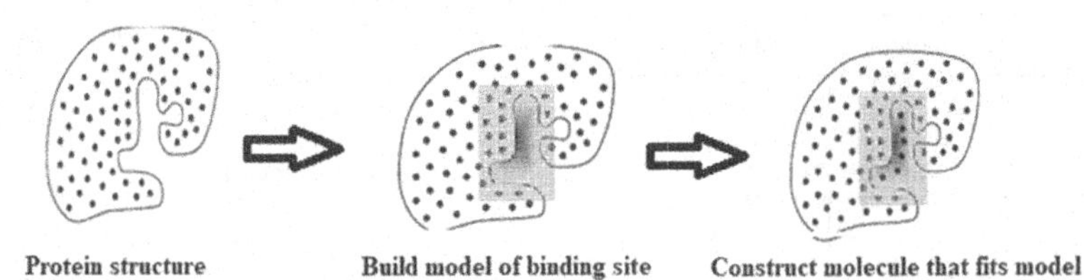

FIGURE 14.5 Demonstration of protein binding.

Design program de novo

- How to assemble the candidate compounds?
- How to determine their potential quality?
- How to sample the search space efficiently?

Figure 14.6 represents the different fragments and other linking groups used in *de novo* design methodology. The fragments linking process described shortly in the following schematic representation.

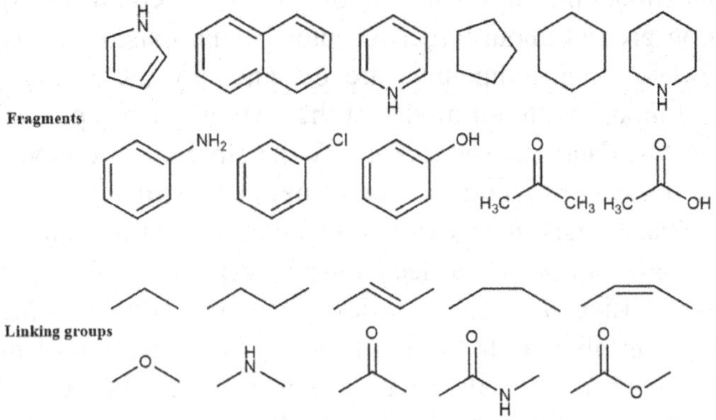

Fragments

Linking groups

FIGURE 14.6 Different fragments and other linking groups used in de-novo drug-design methodology.

14.7.2 PROCEDURE OF DE-NOVO DRUG DESIGN

- To crystallize target protein with bound ligand (e.g., inhibitor + enzyme or ligand)
- To acquire structure by X-ray crystallography
- To identify binding site (region wherever ligand is bound)
- To remove ligand
- To identify potential binding regions within the binding site
- To design a lead compound to relate with the binding site
- To synthesize the lead compound and test it for activity
- To crystallize the lead compound with target protein and to identify the actual binding interactions
- Drug design on the basis of structure (Fig. 14.7)[30]

14.7.3 TYPES OF DE-NOVO DRUG DESIGN

1. Manual design
2. Automated design[31]

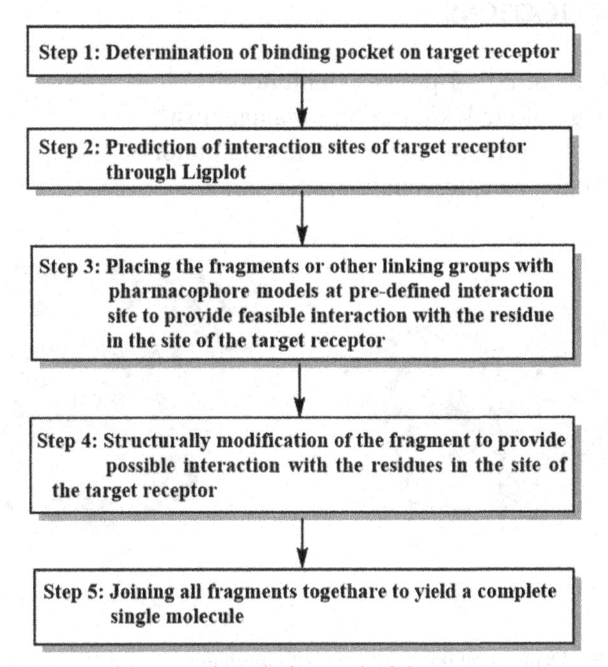

FIGURE 14.7 Steps of de-novo drug-design methodology.

14.7.4 *DISADVANTAGES*

1. The atomic position in the crystal structure is accurate only to 0.2–0.4 A and allowance should be made for the same.
2. It might be possible that the designed molecule not bind to the binding site accurately as predicted.
3. It is worth leaving scope for variation and explanation of the molecule. This allows fine-tuning of the pharmacokinetic and the molecule's binding affinity.[32]

14.7.5 *PROBLEMS OF AUTOMATED DE NOVO*

1. Automated de-novo drug design is used to generate structures that are difficult or not possible to synthesize.
2. Automated de-novo programs revolve about the scoring functions used to guess binding affinities.[33]

14.8 APPLICATIONS

1. Design of HIV 1 protease inhibitors[28]
2. Design of bradykinin receptor antagonist[34]
3. Catechol orthomethyl transferase inhibitors[28,35]
4. Estrogen receptor antagonist[36]

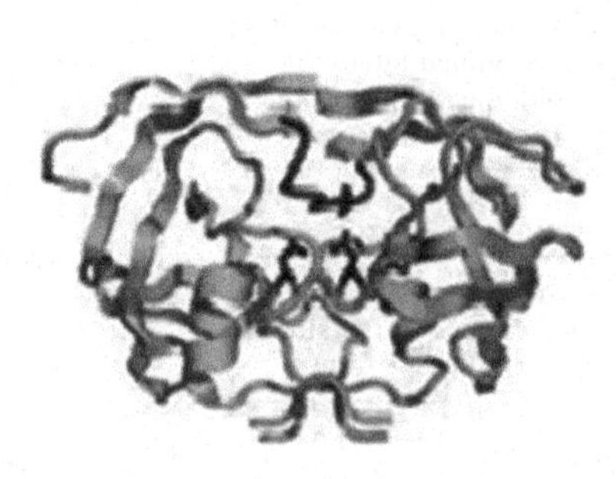

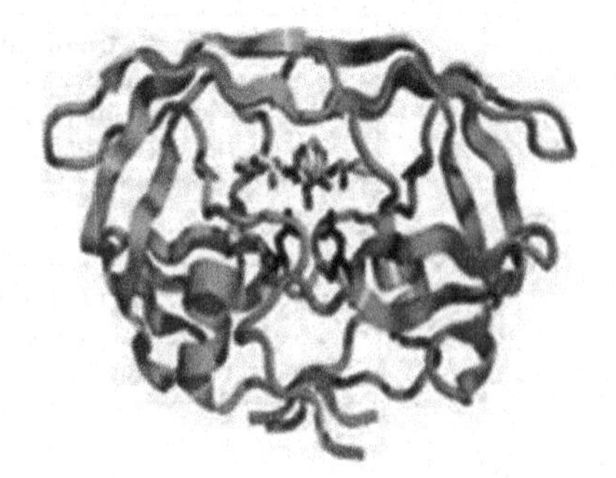

FIGURE 14.8 Demonstration of enzyme with inhibitor.

14.9 OTHERS METHODS FOR DE-NOVO DRUG DESIGN

Method	Programs available
Site point connection method	LUDI
Fragment connection method	SPLICE, NEW LEAD, PRO-LIGAND
Sequential build-up methods	LEGEND, GROW, SPORUT
Random disconnection and connection method	CONCEPTS, CONCERTS, MCDNLG

14.10 CONCLUSION

The goal of molecular docking is to achieve an optimized conformation for both the protein and ligand and relative orientation between protein and ligand so that the free energy of the overall system is minimized. Molecular recognition plays a key role in promoting fundamental biomolecular events such as enzyme–substrate, drug–protein, and drug–nucleic acid interaction. These interactions can be achieved by various methods of molecular docking such rigid docking, flexible docking, and full flexible. Homology modeling is also having importance in prediction of 3D structure of target protein without an experimental structure. In de-novo drug design by using docking programs, new lead structures can be design that fit a particular target site.

KEYWORDS

- **molecular docking**
- **de novo**
- **drug design**
- **computational tool**

REFERENCES

1. Talevi, A. Computer-Aided Drug Design: An Overview. *Computational Drug Discovery and Design*; Springer: Germany, 2018; pp 1–19.
2. Gschwend, D. A.; Good, A. C.; Kuntz, I. D. Molecular Docking Towards Drug Discovery. *J. Mol. Recognit. Interdiscipl. J.* **1996,** *9* (2), 175–186.

3. Zamora, I. Site of Metabolism Predictions: Facts and Experiences. *Anti-targets: Prediction and Prevention of Drug Side Effects*; Wiley-VCH: Weinheim, 2008.
4. Agarwal, S.; Mehrotra, R. An Overview of Molecular Docking. *JSM Chem.* **2016,** *4*, 1042–1045.
5. Shoichet, B. K. Virtual Screening of Chemical Libraries. *Nature* **2004,** *432* (7019), 862.
6. Shoichet, B. K.; Leach, A. R.; Kuntz, I. D. Ligand Solvation in Molecular Docking. *Proteins Struct. Funct. Bioinf.* **1999,** *34* (1), 4–16.
7. Dias, R.; de Azevedo, J.; Walter, F. Molecular Docking Algorithms. *Curr. Drug Targets* **2008,** *9* (12), 1040–1047.
8. Elcock, A. H.; Sept, D.; McCammon, J. A. *Computer Simulation of Protein–Protein Interactions*; ACS Publications, USA, 2001.
9. Dar, A. M.; Mir, S. Molecular Docking: Approaches, Types, Applications and Basic Challenges. *J. Anal. Bioanal. Tech.* **2017**, *8* (02), 356.
10. Mukesh, B.; Rakesh, K. Molecular Docking: A Review. *Int. J. Res. Ayurveda Pharmacy* **2011,** *2* (6), 1746–1751.
11. Tripathi, A.; Bankaitis, V. A. Molecular Docking: From Lock and Key to Combination Lock. *J. Mol. Med. Clin. Appl.* **2017,** *2* (1), 1–19.
12. Nabuurs, S. B.; Wagener, M.; De Vlieg, J. A Flexible Approach to Induced Fit Docking. *J. Med. Chem.* **2007,** *50* (26), 6507–6518.
13. Mizutani, M. Y.; et al. Effective Handling of Induced-fit Motion in Flexible Docking. *Proteins Struct. Funct. Bioinf.* **2006,** *63* (4), 878–891.
14. Moitessier, N.; Therrien, E.; Hanessian, S. A Method for Induced-Fit Docking, Scoring, and Ranking of Flexible Ligands. Application to Peptidic and Pseudopeptidic β-secretase (BACE 1) Inhibitors. *J. Med. Chem.* **2006,** *49* (20), 5885–5894.
15. Bonvin, A. M. Flexible Protein–Protein Docking. *Curr. Opin. Structural Biol.* **2006,** *16* (2), 194–200.
16. Keserû, G. M. Kolossváry, I. Fully Flexible Low-Mode Docking: Application to Induced Fit in HIV Integrase. *J. Am. Chem. Soc.* **2001,** *123* (50), 12708–12709.
17. Zsoldos, Z.; et al. eHiTS: A New Fast, Exhaustive Flexible Ligand Docking System. *J. Mol. Graph. Model.* **2007,** *26* (1), 198–212.
18. Gervasio, F.L., A. Laio, M. Parrinello, Flexible Docking in Solution Using Metadynamics. *J. Am. Chem. Soc.* **2005,** *127* (8), 2600–2607.
19. Nadendla, R. R. Molecular Modeling: A Powerful Tool for Drug Design and Molecular Docking. *Resonance* **2004,** *9* (5), 51–60.
20. Chaudhary, K. K.; Mishra, N. A Review on Molecular Docking: Novel Tool for Drug Discovery. *Databases* **2016,** *3* (4), 1029.
21. de Ruyck, J.; et al. Molecular Docking as a Popular Tool in Drug Design, an In Silico Travel. *Adv. Appl. Bioinf. Chem.* **2016,** *9*, 1.
22. Krieger, E.; Nabuurs, S. B.; Vriend, G. Homology Modeling. *Methods Biochem. Anal.* **2003,** *44*, 509–524.
23. Xiang, Z. Advances in Homology Protein Structure Modeling. *Curr. Protein Pept. Sci.* **2006,** *7* (3), 217–227.
24. Cavasotto, C. N. Phatak, S. S. Homology Modeling in Drug Discovery: Current Trends and Applications. *Drug Discov. Today* **2009,** *14* (13–14), 676–683.

25. Kopp, J.; Schwede, T. Automated Protein Structure Homology Modeling: A Progress Report. *Pharmacogenomics* **2004,** *5* (4), 405–416.
26. Rotstein, S. H.; Murcko, M. A. GenStar: A Method for De Novo Drug Design. *J. Computer-Aided Mol. Des.* **1993,** *7* (1), 23–43.
27. Rotstein, S. H.; Murcko, M. A. GroupBuild: A Fragment-Based Method for De Novo Drug Design. *J. Med. Chem.* **1993,** *36* (12), 1700–1710.
28. Hartenfeller, M.; Schneider, G. Enabling Future Drug Discovery by De Novo Design. *Wiley Interdiscip. Rev. Comput. Mol. Sci.* **2011,** *1* (5), 742–759.
29. Yuan, Y.; Pei, J.; Lai, L. LigBuilder 2: A Practical De Novo Drug Design Approach. *J. Chem. Inf. Model.* **2011,** *51* (5), 1083–1091.
30. Schneider, G.; Fechner, U. Computer-Based De Novo Design of Drug-Like Molecules. *Nat. Rev. Drug Discov.* **2005,** *4* (8), 649.
31. Pearlman, D. A.; Murcko, M. A. Concepts: New Dynamic Algorithm for De Novo Drug Suggestion. *J. Comput. Chem.* **1993,** *14* (10), 1184–1193.
32. Nicolaou, C. A.; Apostolakis, J.; Pattichis, C. S. De Novo Drug Design Using Multiobjective Evolutionary Graphs. *J. Chem. Inf. Model.* **2009,** *49* (2), 295–307.
33. Leach, A. R.; Bryce, R. A.; Robinson, A. J. Synergy between Combinatorial Chemistry and De Novo Design. *J. Mol. Graph. Model.* **2000,** *18* (4–5), 358–367.
34. Schneider, G.; Baringhaus, K.-H. *Molecular Design: Concepts and Applications*; John Wiley & Sons: Hoboken, NJ, 2008.
35. Hartenfeller, M.; Schneider, G. De Novo Drug Design. *Chemoinformatics and Computational Chemical Biology*, Springer: Berlin, 2010; pp 299–323.
36. Clark, D. E.; Pickett, S. D. Computational Methods for the Prediction of 'Drug-Likeness'. *Drug Discov. Today* **2000,** *5* (2), 49–58.

Index

E

N

O

P

Q

R

S

T

V

W

Z

Printed in the United States
by Baker & Taylor Publisher Services

Printed in the United States
by Baker & Taylor Publisher Services